The System of the Sciences

The System of the Sciences
according to Objects and Methods

Paul Tillich

Translated with an Introduction by Paul Wiebe

Lewisburg
Bucknell University Press
London and Toronto: Associated University Presses

Originally published in 1923 as *Das System der Wissenschaften nach Gegenständen und Methoden*
Copyright 1923, Vandenhoeck & Ruprecht

Associated University Presses, Inc.
4 Cornwall Drive
East Brunswick, N.J. 08816

Associated University Presses, Ltd.
69 Fleet Street
London, EC4Y, 1EU, England

Associated University Presses
Toronto, Ontario, Canada M5E 1A7

Library of Congress Cataloging in Publication Data

Tillich, Paul, 1886–1965.
  The system of the sciences according to objects and methods.

  Translation of: Das System der Wissenschaften, nach Gegenständen und Methoden.
  Includes bibliographical references and index.
  1. Science--Philosophy. I. Title.
Q175.T55613   501   80-67078
ISBN 0-8387-5013-3   AACR2

Printed in the United States of America

In Memory of Ernst Troeltsch

Contents

Foreword

The German original of this work, *Das System der Wissenschaften nach Gegenständen und Methoden*, was first published in 1923 (Göttingen: Vandenhoeck & Ruprecht); it was reprinted in the first volume of Tillich's *Gesammelte Werke* (Stuttgart: Evangelisches Verlagswerk, 1959). The present translation is complete; it has been made from the latter edition. I have benefited at various points from two unpublished translations: the very rough translation of the entire work by Emile Grünberg, the first thirty pages of which have been revised by James Luther Adams, and the translation of the "General Foundation" by John D. Poynter and Clark A. Kucheman. The translation is mine, however. The few notes that Tillich included in the text have been placed at the end. One of my own appears there, too, but I have generally held to a strict policy of avoiding editorial intrusion into the body of the text.

Most of the technical terms Tillich uses are staples of German philosophy, though many of them receive the stamp of his own meaning when they enter the orbit of his thought. Something must be said about a few of the basic terms and my translation of them.

The fundamental categories are *Denken*, *Sein*, and *Geist*. *Denken* is rendered as "thought," though it refers both to the cognitive subject of the act of knowledge and to the act itself. *Sein* is "being," the cognitive object; this term should not be confused with the term "being" in Tillich's mature theology, for his early and his later ontologies are not identical, though they are similar. *Seienden* are "existents," entities within the realm of *Sein*. I translate *seinshaft* as "existential," of or relating to being; this English equivalent does not refer to human existence, which is to say that it has nothing to do with existentialism. And *Geist* is "spirit," the sphere of culture and the distinctively human, the sphere in which thought inserts itself into being. No English word corresponds exactly to the German, but I have chosen "spirit" over the other common rendition, "mind": it is less misleading, and Tillich himself preferred "spirit." For Tillich, the act of knowledge is a species of the more general spiritual act, which gives "meaning" (*Sinn*) to cultural life. He uses the pair of terms *Form* and *Gehalt* to refer to the two elements of the spiritual act. "Form" is the element of universality and validity. *Gehalt* is the element of irrationality and infinity; following Adams in his book *Paul Tillich's Philosophy of Culture, Science, and Religion*, I translate it as "import" rather than as "content" or "substance," because this leaves "content" free for *Inhalt* and "substance" for *Substanz*.

Wissenschaft is usually translated as "science," although "science" does not capture the breadth of the German word. I follow this convention, with the explanation that "science" here is intended in the broad sense of "cognitive discipline." The implication is that there are other ways of knowing than the "natural" and the "social" sciences. Tillich outlines three fundamental forms of science, the *Denkwissenschaften*, the *Seinswissenschaften*, and the *Geisteswissenschaften*. The *Denkwissenschaften* are the "sciences of thought," of course, or simply the "thought sciences." Depending on the context, I translate *Seinswissenschaften* as either "sciences of being" or "empirical sciences." *Geisteswissenschaften* has no established translation; one has a choice of several. The German word was originally the translation of John Stuart Mill's "moral sciences," which is now archaic. "Sciences of spirit" is literal, but awkward. "Cultural sciences" is reasonable enough, but it leaves no English word for referring to *Kulturwissenschaften*, which Tillich also mentions but does not include in his system. "Humanities" creates a sense of familiarity that should be avoided; Tillich's *Geisteswissenschaften* simply do not cover the same territory that our humanities do. "Human studies" is becoming quite acceptable, but it weakens the claim, inherent in the German word, that these disciplines are indeed sciences. I have settled on "human sciences"; it is close to "human studies," but avoids the problems of this and the other possibilities.

Finally, *Geistesgeschichte* is "history of spirit," but *Geistesgeschichte des Rechts*, for example, is "spiritual history of law." I simply anglicize *Gestalt*, dropping the upper case and pluralizing in the conventional English way. Tillich assumes a distinction between *Theorie* and *Lehre*; I respect this distinction with "theory" and "doctrine." He contrasts the *autogen* and *heterogen* methods. To avoid a double contrast, I render them as "autogenous" and "heterogenous" (rather than "heterogeneous," the more common word). And to preserve the contrast with psychologism, I follow Adams's coinage of the word "logism" for Tillich's *Logismus*.

I am grateful to Wichita State University for providing research funds that helped me prepare this translation. Individuals have also given me aid, and I am pleased to acknowledge my debts to them. Peter Oliver made the resources of the Tillich Archive of the Andover-Harvard Theological Library available to me. Carl Adamson reviewed my translation of difficult German constructions. John Poynter and Clark Kucheman provided me a copy of their translation of the "General Foundation," and Kucheman gave me assistance along the way, as did John Carey, Alan Anderson, and Anthony Gythiel. Karla Kraft typed the penultimate draft of the manuscipt. Robert Kimball, executor of the Tillich literary estate, kindly gave me permission to publish the project.

I give most special thanks to that great Tillich interpreter, James Luther Adams, who generously gave his time, encouragement, advice, and most gracious aid. It is fitting that this translation is dedicated to him.

—PAUL WIEBE

Wichita State University

Preface

The question of the systematic classification of the sciences has occupied me for years. Work on the individual human sciences continually forced me toward a systematic foundation. I became convinced that a system of the sciences is not only the goal but also the starting point of all knowledge. Only radical empiricism can dispute that; for it, there is no system whatever. But whoever wishes to be scientifically self-conscious (and this is necessary not only for one who works in the human sciences) must be aware of his place, both materially and methodologically, within the totality of knowledge. For every science stands in the service of the one truth, and it perishes if it loses its connection with the whole.

I believe, therefore, that I may venture to publish this outline. I am aware of the audacity of my procedure. In the areas outside my own field, I had to be satisfied with a systematic arrangement and the brief mention of the most general foundations. In the individual empirical sciences, I was able to proceed by treating each science independently. In the human sciences, I submit a fundamentally new systematics. Behind the whole edifice, however, there stands a systematic conviction about the nature and method of the human sciences, and thus of science in general — a conviction that finds summary expression in the "Conclusion."

The limits of such an outline are obvious. The task of overcoming them is the work of an entire generation, not just that of a single life. This book reflects the accidental circumstances of its origin. I first intended to write an introductory textbook on the approaches to philosophy, but this restriction disappeared during the course of the work. Soon I was pursuing two tasks, one constructive and the other pedagogical. This duality, which is still observable in the presentation, kept me from abandoning my original intention completely. It appears to me that it is a duty of the specialist, even if he is philosophically unsophisticated, to prepare a survey of the whole of knowledge and to show the significance of his own area for the entire system of science. I have omitted scientific details, both in accordance with the intention of the book and for the sake of systematic compactness. And it was impossible for me to provide references; had I done so, the book would have assumed unlimited proportions. The presentation itself shows that I considered confrontations with the alternative positions more important than such references.

During the printing, I learned of the sudden death of Ernst Troeltsch. It was his passionate aspiration to arrive at a system. I should like to express the thanks I owe him for the influence his work has had upon the spiritual foundations of this book by dedicating it to his memory.

Berlin-Friedenau, Easter 1923   — PAUL TILLICH

Translator's Introduction

The squirming facts exceed the squamous mind,
If one may say so. And yet relation appears,
A small relation expanding like the shade
Of a cloud on sand, a shape on the side of a hill.

—WALLACE STEVENS[1]

Had Paul Tillich been a poet in art as well as in spirit, he could well have written words such as these. But his craft was philosophy, the architecture of ideas. He did not write small poems; he said what he had to say by composing immense systems—two of them. One of the two, *Systematic Theology*, has found a just fame. It is without doubt an achievement of a very high order. What is not so well known is his earlier system. *The System of the Sciences* was his first large book. When it appeared in 1923, Tillich was already thirty-seven years old. His work on Schelling and Schleiermacher, some of his writings on religious socialism, and a few seminal essays were behind him; his emigration to the United States and his major contributions to Protestant theology were yet to come. In retrospect, we can say that in its own way, this initial system was an achievement, too—not least because it was Tillich's first big step toward the monumental system that was to be completed forty years later.

Despite the anachronism, this early system can be viewed as a sermon on the later Wallace Stevens text. "The squirming facts exceed the squamous mind." Tillich's basic premise is that "being" (the multifarious world) eludes "thought" (the human mind) in infinite degree. "And yet relation appears." Thought is under the unconditioned demand to grasp being perfectly. The sciences show themselves as the many attempts to perform the impossible task that it devolves upon thought to pursue. "A small relation expanding." The simple principle of all knowledge and science yields the formidable arrangement of the sciences. For as the title of the book suggests, *The System of the Sciences* is an attempt to organize the academic disciplines, to trace "a shape on the side of a hill."

The incessant changes in the sciences have often been attended by efforts to order them. For example, universities have continually faced the theoretical and practical problems of organizing their curricula. And encyclopedias have always used some scheme or other for dividing their summaries of knowledge. Tillich's book is neither a blueprint for an academy nor what he calls "a detailed system of knowledge." It places him within the line of those philosophers who have classified the sciences. Plato, Aristotle, Boethius, the Bacons, the French Encyclopedists, Locke, Kant, the German Idealists, Bentham, Comte, Spencer—these and others have constructed schemes for organizing the sciences. Some of these outlines have been the concern of entire works; others have been secondary to the main purpose of the writings in which they have appeared. Some have been based on clear principles of classification; others have been arbitrary or conventional. Some have been predominantly descriptive, seeking to organize existing disciplines; others have been prescriptive or visionary. And some rank as a major accomplishment of the thinker; others are now disregarded. Tillich's system is the major subject of one book, is based upon a considered principle, intends to be prescriptive, and has largely been forgotten.

Tillich is aware of this tradition of encyclopedias: in the early pages of the book he mentions the ancient Greeks, the Platonic school, and Fichte's theory of science. Though he seems to think that all systems of the sciences must follow some classification similar to his own thought/being/spirit scheme ("there is an inner necessity that must continually lead to similar formulations"—p. 37), one suspects that his special debt is to the German Idealists generally (e.g., the logic/nature/spirit triad of Hegel's *Encyclopedia*) and to Fichte in particular. But his own arrangement goes beyond the landmarks of the classical Idealists. It is determined by a discussion that was a prominent feature of the German philosophical landscape of the last part of the nineteenth century and the early decades of the twentieth. Philosophers such as Dilthey, Windelband, and Rickert attempted to define the relationship between the natural sciences and a group of disciplines called the *Geisteswissenschaften*, or human sciences. This term had been coined in the middle of the nineteenth century; it had been used to encompass history, philology, economics, sociology, philosophy, psychology, political science, and the studies of law, religion, literature, art, music, and ethos—disciplines that had developed their own modern methods apart from metaphysics, as the natural sciences had been doing for some centuries. Dilthey and his fellow philosophers all contended that the human sciences are indeed sciences, or disciplined ways of knowing, as clearly as are the natural sciences. Each tried to define the distinctiveness of the human sciences as a whole over against the more established natural sciences. But they debated whether this distinction should be made according to the objects (or subject matter) of the two basic forms of science or according to their methods.

Though he never explicitly says so, it is clear from the treatment itself that Tillich intends his sytem of the sciences to be a contribution to this discussion. He establishes the human sciences as one of the three main forms of science, and the original subtitle of the book shows that he attempts to resolve the dispute over whether object or method is primary by organizing the sciences according to both objects and methods. But he does not accept the fundamental division into the natural and the human sciences. Instead, he organizes the sciences by appealing to the principle of science itself, analyzing the idea of knowledge into the elements of thought, being, and spirit, then deriving the thought (or formal) sciences, the sciences of being (the empirical sciences), and the *Geisteswissenschaften*. And his understanding of the human sciences is different from that of his precursors. He places some of the charter members of the *Geisteswissenschaften* (e.g., psychology, sociology, history, philology) within the empirical sciences and classifies most of the others within the human sciences proper, which he regards as the normative sciences.

But Tillich does not construct his system for purely philosophical reasons. He is primarily a theologian. His motive for formulating a scheme is to find a place for theology within the total framework of the sciences so that he will have a foundation for theological work. Christian theologians before him — Origen, Augustine, Thomas Aquinas, Schleiermacher, for example — were also concerned, in one way or another, with the placement of theology. Unlike most of them, however, Tillich writes at length on the subject, and he constructs a complete system of his own rather than adapting one already at hand.

The historical context of Tillich's effort to place theology is, of course, the progressive secularization of the Western consciousness. This secularization is reflected in many of the modern organizations of the sciences: Kant, Bentham, and Comte omitted theology from their systems altogether. And the prestige of all the new sciences has jeopardized theology's very claim to scientific status. Tillich's response to this situation is to restore to theology the pride of place it once enjoyed — but with two qualifications: that it have preeminence only among the human sciences, and that this preeminence be one in which theology stands, not above the other human sciences, but within them as the normative science of the theonomous attitude.

Yet *The System of the Sciences* has been neglected. In Germany, it was the subject of a few reviews after its appearance, then passed into oblivion. With several exceptions,[2] English-speaking theologians and philosophers have been unaware of its existence, knowing only Tillich's mature, more accessible writings.

The weaknesses of the work — Tillich himself later acknowledged them — have naturally contributed to its neglect. His treatment of those sciences outside his own domain was highly formal and sketchy; the reliance on a limited system of concepts necessarily made him repetitious; his constructions often appeared forced; his readers have accused him of rank obscurity. All these factors have helped prevent *The System of the Sciences* from becoming the classic his *Systematic Theology* has become.

It is ironic that the first book of a thinker who emphasized *kairos*, "the right time," would be so untimely. Yet the fact that this work appeared at just the wrong time in the history of

German thought might be the chief reason it has been ignored. In 1923, the philosophical climate was already beginning to change. The Dilthey/Windelband/Rickert discussion to which Tillich's book was a response was exhausted; the categories of German Idealism—categories Tillich used—were spent. Husserl's *Ideas* had been in print for ten years; Heidegger's *Being and Time* was several years away. The theological situation was also undergoing a decisive transformation. The influence of Barth and his associates was spreading; any concern with the problems of culture and philosophy was at least suspect. So considering its nature, it is not surprising that Tillich's first book was not well received in Germany. And if it was forgotten in Germany, one could not expect it to have been remembered in the Anglo-American world.

The weaknesses and untimely advent of *The System of the Sciences* are certainly not fatal, however. This translation has been made in the conviction that further neglect of the book cannot be justified. There are two audiences to which this resurrected text addresses itself. The first consists of those who seek a finer understanding of Tillich's entire thought; the second, of those who are on the watch for resources to help them think about some problems to which the system might suggest a solution.

1. Judging only by the number of articles and books on the subject, one concludes that scholarship on Tillich's thought has been dominated by those who seem to assume that Tillich began to think and write only after his arrival in America. With the appearance of the present translation and others (James Luther Adams's collection of Tillich's initial writings in the philosophy of religion, Victor Nuovo's work on the Schelling dissertations, Franklin Sherman's translation of the major piece on socialism), this situation is destined to change. It is indeed beginning to change. While I cannot here investigate the relation between the early and later writings in detail,[3] I do wish to make several points about how *The System of the Sciences* can provide a key to a fuller understanding of his theological system.

Tillich intended the placement of theology to be a foundation for his theological work proper. That intention becomes clear when we consider his later development. Immediately after having completed *The System of the Sciences*, where he had established the human sciences as one of the three main forms of science, the science of religion as a constituent of the human sciences, and theology as one of the three elements of the science of religion (together with the philosophy of religion and the spiritual history of religion), Tillich turned to a study of the science of religion. In his next major work, *The Philosophy of Religion* (1925), he treated the basic element of the science of religion. The final sentence of this work pointed ahead, however: he had produced a philosophy of religion that would have a theology as its sequel. So in 1925, he began to work on his theological system. — Now, these are significant facts. Future investigators of Tillich's theology must ask what these facts mean; they must determine the degree to which *Systematic Theology* fulfilled the original intention; they must understand and judge the completed system with this intention — fulfilled or not — in mind.

There are many important continuities between *The System of the Sciences* and the theological system, continuities that make the former an indispensable source for interpreting the latter. Several examples come to mind. The later system of the functions of the human mind is clearly based on the earlier system of the functions of meaning. The theory of critical phenomenology and the various typological constructions found in *Systematic Theology* have their explanation and justification in the doctrine of the relation between the history of spirit and the normative system. The basic self/world ontological structure of the theological system is similar to the thought/being epistemological structure on which the taxonomy of the sciences rests. And the later view that theology, as a normative discipline, uses a norm for interpreting the Christian symbols has its origin and theoretical support in the earlier theory of normative exegesis. In all of these cases, the position of the finished system and the often cryptic arguments used to warrant it require that the interpreter examine their more extensive treatment in *The System of the Sciences*.

2. Potentially more important than the value of the book for contemporary Tillich studies, however, is its significance for present thought. The author of *The System* can be a silent discussion partner in a pair of current conversations: one con-

cerning the very nature of the academic disciplines, another about the place of theology among those disciplines.

In the last two decades, the discussion of the nature of the sciences has entered a new phase, which is defined by a group of theoretical developments that includes a renewed interest in the sociology of knowledge (Merton, Luckmann and Berger), a theory of scientific revolutions (Kuhn), a theory of human understanding (Toulmin), a new hermeneutical theory of the human sciences (Gadamer), an interest theory of knowledge (Habermas), and an archeology of knowledge (Foucault). These various developments have given rise to a series of related questions. What is a science, or academic discipline? How and why does a discipline begin? How does it develop? What social context does it presuppose, and what institutional arrangements does it require? To what degree is knowledge based on extracognitive factors? In what sense does a discipline yield knowledge? Are there really methodological differences between the natural and the human sciences? If so, do the human sciences have a way of gaining objectively valid knowledge? How are the many disciplines related? Is it possible to organize them, or are all attempts to do so bound to be arbitrary?

I will not presume to reconstruct a hypothetical discussion between the young Tillich and our contemporaries about these matters, or to perform the feat of showing that Tillich's theory can resolve all difficulties. It is enough to suggest that his system, perhaps the last statement of the old Dilthey/Windelband/Rickert conversation, has a right to be heard half a century later, in a new conversation, if only because it is an important and creative statement that never received the hearing it deserves.

Tillich's solution to the problem of the place and legitimacy of theology as a science reflects the German situation, in which theological faculties exist within public universities. In America and elsewhere, that problem is defined by a new situation, the emergence of the secular study of religion as a discipline that is located within the pluralistic context of the secular university, which is independent of the theological seminary and thus of ecclesiastical interests. On the one hand, theology has been the normative discipline that states the faith of a particular religious community; on the other hand, the academic study of religion is a discipline that serves no special interests, that only describes how people are and have been religious. This contrast poses a few questions. Can and should theology exist within the framework of the study of religion? If not, how can the study of religion exist if the most important question, the question of norms, is excluded in principle? If so, how can it exist within a pluralistic framework?

Though Tillich's concern with the place of theology was not a response to the present situation, his map of the entire science of religion might be seen as one possible solution to the problem. For Tillich, the study or science of religion would include a normative as well as a descriptive element. Certainly the empirical, descriptive part — what he calls the cultural history of religion — would be indispensable, but it should be completed by the normative part, which has theology at its pinnacle. So Tillich would give theology, as the theonomous science of norms, the central place within religious studies, even though theology is inevitably confessional in the sense that it is always related to some religious tradition. This hypothetical response to a real problem has defects, of course. It is not clear how many confessional forms of theology he would allow, or how these forms would be related, and a theologian in his scheme would be forced to participate in a definite religious tradition in order to construct a religious norm. And in public institutions, the scheme would breach the wall of separation between church and state. This descriptive/normative scheme nevertheless deserves scrutiny by those who reflect on the present situation.

So, what was untimely in 1923 may be timely now, if we adopt the early, forgotten Tillich as a partner for debating some of the vital problems of our day. If this timeliness is recognized, and if students of Tillich use it to see the larger pattern of his thought, this translation will not have been in vain.

The System of the Sciences

General Foundation

1. The Meaning and Value of the System of the Sciences

In order for an object to be known, it must be assigned its necessary place within a context. The individual in its isolation is never an object of knowledge. Where there is no comprehensive context, the individual might be perceived, but it will not be known. This is true for every object and for every science; the sequel will show that, and how, it is true. And it is also true for science itself. For even science is a *fact*. Every particular science is a historical phenomenon: it originates at some point in time from some causes, it is more or less clearly distinguished from the other sciences, and its objects and methods are subject to change. One can therefore seek to understand science as a historical fact; one can examine it in its growth and development; one can write a history of every science and a history of science in general; one can also write a psychology and a sociology—even a biology—of science. In every one of these ways, the sciences would be arranged within a comprehensive context; they would become an object of knowledge.

This arrangement of the sciences would suffice if the sciences were merely facts. But they have another significance as well: they are creations of the spirit, and thus they are subject to the norms and critical judgment of spirit. We ask not only, How many sciences are there, how did they originate, and what methods do they use? We also ask, How many sciences should there be, how should they be structured, and what methods should they employ? In other words, we arrange them not only within an empirical context, but within a *normative* one as well.[1]

But in order to be able to organize and structure the multiplicity of sciences, we must clarify the nature of science itself. Only recently has the scientific spirit raised these questions. Only after the world of objects, on which it had hurled itself with a sense of unlimited power and with boundless naiveté, had repulsed its attack, and only after doubt had destroyed its simple self-confidence — only then did the scientific spirit examine itself and ask, What exactly is knowledge, and what is science? And so it produced the most remarkable and profound of all its creations, the science of science, the *doctrine of science*. Once this task is comprehended, however, the fundamental significance of the doctrine of science for everything else becomes evident. What was last in the order of time is first in the order of things. We will consider the systematic placement of the doctrine of science in its proper place. But one of its tasks — the one, indeed, that is ultimate and most comprehensive — is to construct the system of the sciences.

Here we confront a question. Is it not most unprofitable to be concerned with a science of the sciences? Is this not a mere formalism, which is able to stimulate the interest of scholastic minds but is meaningless for living knowledge? Is not knowledge itself better than the knowledge of knowledge, which must then itself be known, and so on to infinity and futility? Is it not better to pursue science itself than to create a system that the living process of knowledge continually renders obsolete?

This objection requires a reply. Even granted that the problems we face are formal, their solution would still give the spirit immense satisfaction. For to solve them would fulfill the spirit's highest desire, which is to apprehend the *unity* of all individuals. How can spirit attain this unity? We can no longer believe that it is possible to achieve this unity through a detailed system of all knowledge, a system that would include the cognition of every individual. Such an attempt is not only extrinsically impossible, it is intrinsically so. The inner infinity of every existent always prevents the completion of this kind of construction. A detailed system of knowledge is an ideal, not a conceivable reality. Yet what is both possible and necessary for every generation is a formal system of knowledge in which one becomes conscious of the realm governed by the spirit, of the objects within this realm, and of the way in which the spirit governs this realm (i.e., the methods). The desire of the spirit to achieve a living unity of knowledge produces this edifice of the system of the sciences, an edifice that appears to be so formal and empty. The power and vitality of the spirit are manifest not in its extensive knowledge of details, but in its ability to unify this knowledge. And although the material of knowledge remains a strange, inexhaustible ocean, form is spirit's native element; after periods of attention to this material, spirit must repeatedly become conscious of form, each time in a richer and more penetrating way.

Even knowledge itself demands the construction of a system of knowledge, thus refuting the reproach that a system is an empty formalism. The organization of the sciences depends, first of all, on the variety of *objects of knowledge*. It is of the highest material significance to discover how the different branches of science are distinguished from each other, which of them are independent and which not, and how they are arranged and systematized into a hierarchy. By answering such questions, the system of the sciences acquires the power to formulate concepts; it is thus not merely a scholastic play with previously established concepts.

The significance of the system of the sciences becomes even clearer when the system is considered in the light of the presuppositions of Kant's epistemology, i.e., when the categories of objects are interpreted as the basic functions of consciousness. Then the system of the sciences becomes the expression of the system of the *functions of spirit*, and the structure of spirit becomes discernible from the different directions in which science locates and delineates its objects. The structure in the subjective realm is apprehended by means of the structure within the objective realm.

The *methods* by which knowledge grasps objects correspond to these objects themselves. Therefore, the most important task of the system of the sciences is to determine the relation between methods and objects. It is, of course, wrong to assume that to every object there corresponds a method appropriate to that object and to that object alone; rather, the same method is applicable to a variety of objects, and the same object is receptive to many methods. Indeed, closer scrutiny shows that the distinction between object and method is fluid. This raises the major problem of determining the degree to which methods create objects. All these questions are answered by the system of the sciences, which must indeed be a system "according to objects and methods."

Thus the system of the sciences becomes the arbiter in the conflict among methods over the same object; it determines boundaries, but also establishes the right to transgress these boundaries. It restrains the unjustified imperialistic claims of particular sciences and methods; it brings hidden possibilities to light, and it shows that different sciences and methods have a right to cooperate; it therefore not only classifies, but also provides guidance. A system of the sciences is a necessary act of scientific self-consciousness for every era. It prevents the sciences, each of which seeks to arrange every individual belonging to its sphere within its own comprehensive context, from standing alongside and in competition with each other in a disordered, chaotic way. It is concerned with form, but it is not formalistic; it is, rather, the living and therefore constantly changing expression of the scientific consciousness of an era.

It is interesting to trace the changes in the sciences and in their methods throughout the course of history: to show how the spheres of knowledge and the moral life are associated with nature, the oldest and most immediate object of scientific investigation; how at the same time the consciousness of the mathematical and logical necessities provides spirit with the most powerful impulses; how all of this becomes material for religious thought; how historical thought emerges from sources completely different from this line of development, and again how psychological thought, reflection on the ego, appears from still other sources; how the multitude of technical sciences grows out of the pure sciences and strives for equal status; how, in modern times, society and economics create new objects as well as new, imperialistic methods; how the struggle for supremacy within the sciences rages between ethics and the philosophy of nature, between theology and metaphysics, and how epistemology briefly attains the position of highest dignity; and how systematics is thus continually impelled to new efforts — efforts that, in spite of all change, contain common basic elements and that establish a significant unity between the systematics of the ancient Greeks and the present one. The presentation of such a spiritual history of the sciences is beyond the scope of this book. A history of this sort must be distinguished from a normative presentation, even though every normative system must emerge from the history of spirit. We will therefore proceed by constructing the system on the basis of a single principle, using and transforming the historical material in particular cases.

But when a genuine system is created (and this is the final response to the accusation of formalism), something more than mere form is manifest in the form. The living power of a system is its *import*, its creative standpoint, its original intuition. Every system lives from the principle on which it is based and by means of which it is constructed. Every ultimate principle is, however, the expression of an ultimate intuition of reality, a fundamental attitude toward life. Thus in every moment, an import breaks through the formal system of the sciences, an import that is metaphysical — that is, an import that lies beyond every particular form as well as beyond the totality of all forms and that therefore can never itself be one form alongside other forms, as it is in a pseudometaphysics. The metaphysical element is the living power, the meaning and the lifeblood, of the system. In this sense, but only in this sense, the formal system of the sciences is metaphysical.

2. The Principle of the System of the Sciences

Every systematic classification of the sciences must proceed from one *principle* that can only be the essence of science itself.[2] Though this axiom is indubitable, it has often been neglected. Past classifications have proceeded less from a strictly systematic interest than from the practical need to create a general view of the extant disciplines and to group the related disciplines and to separate the unrelated ones for the purpose of working out a division of labor. Even today, the division of the faculties and the differences among institutions of higher learning obscure both the essential relatedness of the sciences and the unity of science that lies behind them all.

If one establishes a *perspective* for classification, one has not yet found a principle. There is a whole series of such perspectives: classification according to the relation of thought to its object (the formal and the empirical sciences), according to the kind of object (the natural and the human sciences), and according to the method used (the natural and the cultural sciences), to name only the most important. A conception of science is implicit in all these schemes of classification, but it has not been elevated to the level of a principle. The perspective is adopted arbitrarily and therefore competes with others that have been chosen just as fortuitously. Hence those who construct such schemes find it necessary to supplement one mode of classification with another, without clarifying the relation between the two. In order to achieve a genuine systematic structure, we will attempt to penetrate behind the perspectives to principles, and behind the principles to *one* principle. A principle is always both a point of departure and a continuation, a foundation and a regulative idea. Such a principle for a system of the sciences can be found only in the idea of knowledge itself.

In order to grasp the idea of *knowledge*, it is necessary to abstract completely from everything objective, as well as from all psychic contents, and to attend to the pure meaning of that which is implied in the essence of knowledge. Every act of knowing contains two elements: the act itself and that toward which the act is directed, or the intention and that which is intended. If we refer to the act by which consciousness directs itself at something for the sake of grasping it objectively as *"thought,"* and if we refer to that toward which the act is directed as *"being,"* we have distinguished thought and being as the two basic elements of knowledge. "Thought" in this sense should not be confused with the psychological act of reflection. Reflection, or the entire psychological process of thinking, is only *one* manifestation of thought: its existence in a living individual. But we must disregard all these relations, for here we are concerned with the pure essence of knowledge, paying no attention to how it occurs and where it exists. The same thing is true of the concept "being"; it should not be confused with the conceptions of an existing thing or an existing substance. These are notions in which thought and being have already been united; they are already contents of knowledge. Here we are concerned with the ultimate elements of knowledge. Thus we can define thought only as the act that is directed toward being, and we can define being only as that which is intended by thought, as that toward which the act of thought is directed. It is utterly impossible to transcend this reciprocal definition of the fundamental concepts, for every higher concept would again be an idea of an existent and would thus contain both elements. Consequently, neither thought nor being can be defined, properly speaking; each must be viewed from the standpoint of the other.

Thought is the act that is directed toward being. But what is the nature of this directedness, what is the relationship between the two elements present in the cognitive act? This relationship can be described in three propositions.

1. Thought posits being as that which is comprehended or conceived, as that which is determined by thought.

2. Thought seeks being as that which is strange and incomprehensible, as that which resists thought.

3. Thought is present to itself in the act of thought; it is directed toward itself and thus makes itself an existent.

The first proposition can read: being is thought-determination (*the principle of absolute thought*). The second can read: being contradicts thought (*the principle of absolute being*). And the third: thought is itself being (*the principle of spirit*).

The fundamental organization of the system of the sciences is based upon these three propositions.

1. Thought attempts to absorb every object—whether it be a part of nature, a historical figure, an emotion, a social form—into the sphere of complete consciousness with the aid of concepts, laws, and contexts into which it inserts the object, until there is nothing obscure or strange remaining. But concepts, laws, and contexts are creations of the thought present in the knowing process. Thus thought dissolves the whole of reality into a network of determinations, until all being is cap-

tured in the unity of thought and therefore being is itself dissolved in thought.

2. But every instance of thought involves more than mere thought itself. There is something else involved, something that lies beyond every process of thought, something that in itself "is" apart from any consciousness of it, something that stands before every consciousness as indissoluble, something that cannot be totally absorbed, that must simply be acknowledged. This is the strangeness of being to thought; it is the infinite gap between the two, a gap that is nevertheless overcome again and again whenever thought plunges into being and seeks to consume being in thought's own pure determinations. Every act of thought, every instance of consciousness, even the entire conscious process of life, contains this conflict between the unity and the strangeness of the basic elements of knowledge. This almost sounds like a myth in which the concepts "thought" and "being" are elevated to the status of primordial divinities whose titanic struggle explains the origin of the world. This is not at all what we mean. What we have in mind is the meaning inherent in every conscious act, however simple. In seeing this color, in recognizing these human beings, in examining these plants, in remembering this battle — in every comprehension of an object, this "divine struggle" is present. Therefore everyone can observe this struggle within himself; indeed, it can only be known through introspection. This is the meaning of the two propositions about the unity and the conflict of thought and being.

3. To these two propositions, however, we must add a third: namely, the remarkable fact that thought is directed not only toward being, but also toward itself — that it observes itself, as it were, while it thinks. In this way, it makes itself into an object alongside other objects. Thought subjects itself to all the conditions and determinations that apply to being and into which thought has dissolved being. Thought becomes a part of existence. If we ask where this existing thought is found, we can only answer: in the "interior" of the conscious being, and for humans, above all, in the spiritual life of humanity.

Thus the self-examination of knowledge has led us to three

basic concepts: pure thought, pure being, and spirit as existing, living thought.

These concepts recur in diverse formulations throughout the history of philosophy. They appear at their sharpest, perhaps, in Fichte's theory of science, which can be understood only if it is interpreted, not as a fantastic metaphysical speculation, but as a self-examination of living knowledge. There is an inner necessity that must continually lead to similar formulations. We are therefore justified in making these three elements the foundation of the system of the sciences: (1) the pure act of thought, (2) that which is intended by this act and thus transcends it, and (3) the actual process in which thought comes to conscious existence — in other words, the triad of thought, being, and spirit.

3. The Structure of the System of the Sciences

It might appear that the concepts we have developed are completely unsuitable for structuring the system of the sciences. For what does it mean to speak of a science of pure thought? Is that not something completely empty? Or what is a science of pure being? How can one say anything at all about a pure being, that is, about something containing absolutely no determinations by thought?

Of course, one can develop a science of thought only if one refers to thought insofar as it is directed toward being; conversely, one can speak of being scientifically only if one presupposes that it is determined by thought. So there is a fundamental distinction between these two groups of sciences, not just a simple contrast. In the *sciences of thought*, cognition is directed toward thought insofar as thought is abstracted from every definite content; cognition is directed toward the universal forms to which every content must be adapted, simply because they are the forms of thought itself. In the *sciences of being*, the contents appear and compel thought to adapt itself to them, naturally within the limits of the forms inherent in thought itself. This is the most general distinction between the sciences of thought and those of being. In the thought sciences, thought is confined to itself; in the sciences of being, it emerges from itself and surrenders to being. In the former, thought is related to itself through the self-intuition of its own pure forms, but in the latter, thought grasps the "other," the "object."

The third element of knowledge is spirit, or thought as a phenomenon, as a being alongside other beings. The third group of sciences is therefore constituted by the *sciences of spirit*, the "human sciences." This notion must immediately be defined in order to establish its position as one of equal status alongside the other two basic groups. For attempts are often made to include the human sciences within one of the two other groups. It is maintained that thought as existing, as conscious spiritual life, is one being alongside others and must therefore be examined by the methods of the sciences of being. This is said, for example, whenever psychology is regarded as the fundamental human science. To say this, however, is to forget that when it thinks about itself, thought does not merely observe itself, as it does all other being; rather, while it is doing this, it determines and criticizes itself, giving itself norms. Its own existence is not strange and remote to thought, as are the existence of a stone or an optical law. When thought becomes conscious of itself, it can never remain merely a disinterested spectator; it is always simultaneously a participant. The human sciences are productive. In them, thought is creative and gives laws.

Just as it is incorrect to classify the human sciences among the sciences of being with the aid of psychology, so it is equally wrong to classify them among the thought sciences with the help of logic. The latter has also been attempted. But such a position overlooks the fact that the distinctive mark of spiritual life is precisely its significant affiliation with being; it forgets that every act of genuine spiritual life realizes more than a logical form, that an irrational givenness that is initially strange to everything logical breaks through, uniting itself with the logical and thereby becoming spirit, never just a form of thought. All varieties of logism forget the irrationally creative character possessed by the life of spirit by virtue of the fact that spirit is fulfilled within being. The creative element is the peculiar feature that distinguishes the spiritual process from both mere being and the mere form of thought. If we use the term "norm" to refer to the creative moment in the life of spirit, we must posit the human or normative sciences as a third group of sciences.

This introduction cannot proceed beyond these brief indications. The details and exact analyses appear in the separate sections of this book; only after we have discussed them can a concluding synopsis fully clarify the relationships among the three main parts. The fundamental division given here is closely related to the first major division of the sciences—into logic, physics, and ethics—that was customary in the Platonic school: logic as the science of pure form, physics as the science of objective reality, and ethics as the science of norms. This triad corresponds to the triad of the *ideal*, the *empirical*, and the *normative sciences*, in spite of all the significant differences in detail.

4. The Method of the System of the Sciences

The method of the systematics in the sphere of science is dependent upon the method of the human sciences in general. Therefore it can be established only when we treat those sciences. It is necessary, however, to anticipate that presentation by making a few important points, in order to prevent a misunderstanding of the basic foundation, as well as of the development, of the entire system. The method of the human sciences is metalogical; logical because of the forms of thought, metalogical because of the import of being. But the two constitute a unity. Mere logism does not do justice to the import of being, and alogism does not do justice to the forms of thought. The former leads to formalism, the latter to arbitrariness. Logism violates all the sciences by forcing them into a formal, logical schema, and alogism is incapable of producing a systematic structure that is self-contained and intrinsically necessary. The method of systematics must thus include both elements. This is possible only when being is not regarded merely as a logical category, but is also perceived as a living import. Indeed, being can be approached through the aesthetic, ethical, social, and religious functions as well as through the logical function. For each of these functions, being is something different, yet in all of them, the same being is intended: the unconditionally real that gives import to all forms. Now, the task of logical thought is to allow these approaches to being to operate; its task is to find forms that express (without impairing their logical correctness) the fulfillment with the existential import that is grasped by all these functions. We call this method "metalogical" in conscious analogy to the term "metaphysical."

The justification for such a method is that thought in its pure self-comprehension (our starting point) has grasped its difference even from the act of knowing, which is only one spiritual reality alongside others. From this follows the fundamental distinction between thought, as a pure act of directedness toward being, and reflection, as one specific spiritual realization of this directedness, the realization in which the formal element achieves purest expression. In addition to reflection, however, there are other acts directed more to the element of import, such as the acts of intuiting, shaping, and believing. The essence of the metalogical method is that it projects the irrational element of these functions into that which is logical. In this way, the concepts "thought" and "being" receive a thoroughly metalogical tone: "thought" becomes equivalent to "form in general," and "being" to "import in general." "Thought" expresses the rational, shaping, form-bearing element, and "being" expresses the irrational, vital, infinite element, the depth and the creative power of everything real. If this explanation is disregarded, the use of these two concepts in all our expositions will be unintelligible. These concepts are not only logical categories, they are metalogical categories as well.

One more explanation will show how the metalogical method operates. Pure logism attempts to grasp all that is real by means of an unchanging rational form. It is necessarily static, because the highest rationality is the unchanging identity of the equation "$A = A$." But the metalogical method is *dynamic*. It reveals the import through the living movement of form, a movement that does not escape the logical unity. In its inception, Hegel's dialectic was a dynamic method. Despite all the criticisms and rejections it has encountered, Hegel's dialectic was the last great system-creating method to exert an influence, whether this influence be conscious or unconscious. Its weakness was that the logical element in it devoured the metalogical and dynamic elements, that at a certain point in the logical and temporal development, the dynamic element was abolished. Thought wished to subdue being completely, but being cannot be subdued, either in part or as a whole; being is intrinsically infinite and possesses creative power. This is why the metalogical method is always a dynamic method as well.

In a dynamic view such as ours, the system of the sciences is envisaged as a living contradiction and a living unity of thought and being. Whenever we use these two concepts, however, we mean them in neither the merely logical nor the mythological senses. They are not empty categories, and they are certainly not primordial, mythological divinities. They are the elements of meaning within reality; they are grasped by all the functions of spirit and presented in the form of scientific concepts.

PART 1

The Sciences of Thought (The Ideal Sciences)

1. *Foundation*

In the thought sciences, scientific knowledge is directed toward those forms that are essential to thought; in so doing, it disregards the connection between thought and being, though it is aware of the possibility of this connection. Two sciences possess these characteristics, *logic* and *mathematics*. The relationship of the ideal sciences to the human sciences and to the sciences of being is reflected in these two sciences: logic examines thought in abstraction from every content, but in considering its objects, it assumes that it is possible for thought to grasp these contents; mathematics likewise examines thought in abstraction from every content, but in considering its objects, it assumes the possibility of the existence of a reality that is full of content. Therefore, logic seeks to know the conditions resident in thought under which objects can be grasped by thought, and mathematics attempts to know the conditions residing in thought under which the objects can exist. Mathematics is the older of the two sciences. For thought is directed first of all and immediately to being, only then to itself. But according to rank and order, logic enjoys a priority, because logic is the presupposition of mathematics, not conversely.

Every object requires a definite attitude on the part of the knowing subject. We call this attitude *method* in the broadest sense. For systematics it is just as important to know the relationship among methods as it is to know that among objects, because the structure of the system is determined equally by both. The discussion of the relation between objects and methods belongs within the section on the "Foundation" of the sciences of being. Here we shall discuss the nature and the various aspects of the methodological problem in general, as well as the method of the thought sciences in particular. Method in the broadest sense has four interrelated sides, which change with the alteration of the fundamental attitude.

1. The *goal* of knowledge: the kind of conceptual formulation within every group of sciences.

2. The *attitude* or *position* of knowledge: the relation between the knowing subject and the object known.

3. The *procedure* of knowledge: the modus operandi, or the method in the narrower sense.

4. The *degree* of knowledge: the kind of certainty that can be attained by a specific procedure.

We will consider these four sides in the "Foundation" of every group of sciences.

In every formulation of concepts, the goal of knowledge is that of grasping being. But being can be grasped only insofar as it is formed by thought. Pure being is essentially ungraspable; it is the abyss of knowledge. Only being that has been formed by thought is an object of knowledge. Scientific formulation of concepts must thus be adequate to the formation of being that is prevalent in each group of objects. Now, this statement appears to contradict the situation within both the sciences of thought and the human sciences. Of course, in neither case is knowledge directed toward being; it is directed toward thought itself, in the one case as pure form, in the other as norm. This is the problem of the ideal and normative formulation of concepts. It is directed toward existents that are not existents—a paradox for which logical terminology has introduced the concept "*validity*." Validity cannot be more precisely defined, because every definition would have to use concepts of either being or validity. Validity is a primitive function. All it means is that thought attempts to realize the unconditionality of its form in every existent, but that no existent corresponds to the pure form. The concept "validity" means that the form of thought stands over against everything real as both formation and demand, in accordance with the basic relationship between thought and being. The concepts of the ideal sciences are thus concepts of validity. But how should we characterize these concepts? We call a completely formed, self-contained existent a gestalt. The concept "gestalt" occupies the central position within the sciences of being, although it is by no means confined to this central group. For both above and below the gestalt there are incomplete gestalts, which we will call "law" and "sequence." Here we must ask whether and in what sense these two concepts are also applicable to the ideal sciences.

Doubtless one can call every logical and mathematical form a "gestalt" or a "law," though the term "sequence" (which refers to the temporal, historical factor) is by its very nature inapplicable to such forms, for the pure forms of thought are unrelated to time. One might say that logical and mathematical concepts are concepts of ideal gestalts and their structural laws. But ideal gestalts are not gestalts in the proper sense, and ideal laws are not really laws. Both concepts are used in the thought sciences only by way of analogy. A genuine law controls all individual events that are determined by it. For a mathematical law, however, there are no individual cases: there is only the individual case of the ideal structure that is determined by the definition of this case. Therefore, one ordinarily speaks of logical-mathematical "propositions." In a genuine gestalt, there is a living functional organization that is distinct from every other gestalt. In comparison, logical and mathematical gestalts are not formed from within; they are formed from the outside, by means of definition; what is more, they are interchangeable. It is thus better to call them "*structures*" than "gestalts." "Structure" indicates the absence of the existential element. The formulation of concepts in logic and mathematics is therefore concerned with structures and propositions about the nature of these structures. Both of these concepts (i.e., "structure" and "proposition") indicate the difference, as well as the relationship, between the goal of knowledge in the thought sciences and that in the sciences of being.[1]

The attitude of knowledge corresponds to the goal of

knowledge. Since thought remains within itself in the ideal
sciences, knowledge can attend only to the laws inherent in
thought, laws it obeys in the cognitive process. We call this attitude, in which knowledge examines its own immanent forms,
"intuition." The attitude of the thought sciences is intuitive.
The methodological concept "intuition" must be distinguished
from its psychological counterpart. In all knowledge, whether
it is scientific or practical, immediate apprehension is decisive,
either alongside or above reflection. But that is still not
methodological intuition. The latter is only present when the
nature of the object itself requires an intuitive attitude, that is,
when the object of knowledge is present in the process of
knowledge itself.

But the term "intuition" does not sufficiently describe the attitude peculiar to the thought sciences. There are still many
propositions and structures that are related to each other by
reflection rather than by intuition. The *rational* attitude of
knowledge is closely connected to the intuitive one. Pure intuition would either remain within the realm of axioms or it
would erect a series of disconnected propositions. There is no
science without *ratio*, without connected conceptions. In the
ideal sciences, however, *ratio* is confined to that which is accessible to intuition. In these sciences there is, of course,
nothing empirical, nothing existential, nothing strange to
thought. *Ratio* is always based upon intuition. Its only foundation is the intuitive certainty of axioms and definitions.

To the intuitive and rational attitude of knowledge there
corresponds the *demonstrative and deductive* procedure of
knowledge. Deduction develops the demonstratively given
structures according to their inherent organization. In the
deductive method, reason is completely autonomous—the
ideal of all rationalism. This method has one presupposition,
however, that is not deducible, one foundation of all deduction. This presupposition is accessible only to demonstration; it
cannot be proved, it can only be indicated.

Finally, the degree of knowledge corresponds to the method:
the intuitive, rational attitude creates *self-evident* knowledge.
The infinite distance between thought and being does not exist
here, because thought is directed to itself, so there is no prob-

ability, no more or less; there is only certainty. Every logical or mathematical proposition expresses the unconditionality of the pure form of thought; that is its dignity, but that is also the reason for its strangeness from real being.[2]

2. Logic

It is not our task to investigate the methodologies of the various sciences in detail. In this system, our main concern is only to establish the conception that is immediately given by the position of every science within the system of the sciences. When we classify logic as a thought science, we must define its differences from both the sciences of being and the human sciences. This task is necessary, because our conception of logic is not universally accepted.

Our conception can be disputed from two sides: logic can be regarded as a science of being, and it can be considered a human science. The first position interprets logical thought as an empirical phenomenon that, like every other phenomenon, must be grasped by perception. For this position, logic has no validity beyond the observable sphere of actual human thought, nor can it attain absolute self-evidence and intrinsic certainty. This point of view is defended most strongly by those who make psychology the foundation of logic, who conceive of logical laws as the consequences of psychic regularities. For them, the certainty of logic is accordingly based solely on the natural compulsion that the psychic organism exerts upon thought. *The compulsion of nature*, not the validity of truth, makes logical laws self-evident. And it is entirely possible that this disposition might change in the course of time, or that there might be different creatures with a different disposition and thus with a different logic. The most that is attainable is a logic of the human race, not a universal logic; a valuable logic, not a valid one.

This position can be criticized in the following way. The knowledge of the psychic processes that lead to a judgment, a principle, or a conclusion tells us nothing about the truth or falsity of the judgment, principle, or conclusion. Even if I know how the mind has arrived at the proposition "A = A," I still know absolutely nothing about its meaning and validity.

But in logic, it is the latter alone that matters, not the origin of the proposition in someone's mind. Logic seeks to comprehend the *meaning* of the proposition, not its existence. But if this is true, every limitation of logic to an era, or to a class of beings such as the "human race," also ceases.

Every such limitation is possible, of course, only by means of logical laws to which both the one making the limitation and that which is limited are subject. If one speaks of another period, or of other minds, with another logic, one in any case presupposes that the principles of identity and the excluded middle are valid for the present period as for every other period, for man as for every other kind of spiritual creature. For if these principles were not valid, every distinction would be meaningless, everything would merge into chaos, and no form could be distinguished from another. Every restriction of the universal validity of logic leads to chaos, drawing into this chaos even the ground supporting the validity of logic. The unconditioned self-certainty, the absolute self-evidence, of logic is also based upon this foundation. One can conceive of changes in every accidental, merely given disposition without lapsing into meaninglessness. It is different with logic. The abolition of logic would mean the abolition of thought itself. It is wrong to say that thought is independently given and that therefore logic's modes of functioning must accompany it (in a way similar to the way in which seeing is directly mediated through the function of the eye, though it could also be mediated in a different way); rather, thought and the laws of thought are one, and therefore, the abolition of the laws of thought would mean the abolition of thought itself. This is why logic is self-evident, and it is why logic, as the science of absolutely valid form, precedes every other science.

Just as it is incorrect to place logic within the sciences of being, so it is wrong to classify it among the human sciences. This usually happens by reducing logic to epistemology. The idealistic disciples of Kant, especially, are inclined to do this. They protest against *formal* logic, thus approaching the view taken by Aristotle, the creator of logic. For him, too, logic was the expression of a definite metaphysical and epistemological point of view. Certainly it is both possible and necessary to interpret logical laws in a metaphysical way. But there can be no justification for founding logic upon metaphysics and reducing it to epistemology, because the formulation of any principle and the interpretation of a logical proposition in terms of that principle already presuppose the validity of logical laws. This is the permanent justification of formal logic, which cannot even presuppose either the distinction between the knowing subject and the object known or the existence of a world in need of explanation. Therefore, everything depends upon the formal nature of logic; because of this formality, logic is the science of pure thought, distinct from both the comprehension of being, which must always be incomplete, and the concrete, individual human sciences.

We must also reject the attempt to regard logic as a *doctrine of methods*. This attempt is closely related to the conception of logic just discussed and is subject to the same criticism. No doubt it is possible to inquire into the significance of the doctrine of inference for knowledge, for example, and perhaps to debate the value or worthlessness of the syllogism for science; but these are not logical problems, they are epistemological ones, and the validity of the rules of inference is independent of them. Validity and usefulness are completely different; their separation confirms the formal character of logic.

Finally, the attempt is sometimes made to consider logic a *doctrine of the art* of thought. But this is the conception that causes the widespread antipathy to formal logic. And it is not clear what scientific results such a conception is supposed to yield. The laws of logic can be derived only from their intrinsic self-evidence. Though to append a "thou shalt" and a "thou shalt not" to the laws discovered in this way may give them a practical, though limited, usefulness, this is certainly not a scientific achievement. Nothing has damaged the reputation and dignity of pure logic more than this academic formalism, which appears with its commands and provokes opposition instead of manifesting the inherent sublimity of the eternal laws of thought that govern the world.

3. Mathematics

The objects of logical propositions are taken from the relation to the process of knowledge; but the content of mathematical thought is determined by its relation to spatiotemporal existence. On the other hand, just as the validity of logical propositions is completely independent of the actual process of thought, so mathematical propositions are independent of the actual existence of things. Mathematics is therefore as much a science of pure form as is logic. Indeed, mathematics is that science in which spirit, both in ancient and in modern times, first became aware of formal validity, the intrinsic self-certainty of thought. It has produced more than logic has and has moved the human spirit more profoundly. The thrill of the mystery of numbers runs from the most archaic number magic to the amazing symbolism of the irrational and imaginary numbers.

Objections to our systematic classification of mathematics can again be raised from two sides: from the side of the sciences of being and from that of the human sciences. The first objection focuses upon *geometry*, the second on *arithmetic*. Nothing seems clearer than that geometry attempts to measure the actual relations of spatial things and to represent these relations by numbers. The whole work of geometry appears to be meaningless if it does not attain this goal. But it is obvious that geometrical structures in no way coincide with actual things, that nature contains no actual points, straight lines, right-angled triangles, or other geometrical structures. And geometry will never be in a position to grasp reality completely. Every line drawn by geometry is determined in every section, however infinitely small, by the law of the entire line; in nature, however, not a single section is precisely determined by this law. The distinction between geometrical concept and reality of nature is absolute and qualitative; it cannot be overcome by any approximation.

Geometrical self-evidence therefore never depends upon *measurement*. The lines that are drawn for the purpose of elucidation are not proofs; they are visual representations of meaning. It is thus foolish even to consider the possibility that measurements will some day be able to show the incorrectness of any geometrical propositions. The geometrical objects that would yield this result would certainly be different from the structures defined by geometry proper, and they would have to be calculated on the basis of other definitions. But the self-evidence would be the same in both cases.

The conception of geometry as an empirical science cannot claim that *space* is a natural object open to empirical investigation. If it does, the question arises: Which space? Certainly the space of geometry is not the space of perception, which originates through the sensations of sight, touch, and motion and which is different with every shade, color, and movement and for every eye. Rather, the space referred to is homogeneous geometrical space. That is to say, geometry creates the space it seeks to calculate; space itself is its first axiom. This state of affairs could remain hidden as long as Euclidean space was the unquestioned presupposition of geometry. But ever since the advent of non-Euclidean geometry, it has been clear that geometry is a science of pure form and that its self-evidence rests solely on the fact that in considering the structures of geometry, thought is autonomous. Hence it is possible to abolish all axioms of the naive geometry that is apparently shaped by empirical reality, and yet not only to retain self-evidence, but also to establish and justify more profoundly the view that the geometry concerned with the world of nature is a special case of geometry as such.

The same objection to our classification applies to arithmetic. One can also regard arithmetic as an empirical science; one can explain *number* as the result of the counting of actual objects and accordingly envisage the day when one might arrive at different results by recounting. This explanation could be meaningful only for the natural numbers, of course; it would not apply to the others, the fractions and negative numbers, the irrational and imaginary ones. But it does not even apply to the natural numbers, for there are no homogeneous things in nature, just as there are no homogeneous lines. But the numerical series presupposes absolute homogeneity. Thus this series itself creates the presupposition it requires. The counting of actual objects is a means of demonstration, not a basis for certainty. Consequently arithmetic, too, is excluded from the sciences of being.

Yet neither arithmetic nor geometry is a *creative and*

aesthetic function, as could perhaps be maintained from the side of the human sciences. The pure form of thought itself is present in every arithmetical and geometrical form; because of this, every one of these formulas can be useful for the knowledge of reality. The self-contained world of the ideal numerical and spatial structures is not directly related to reality. Since every ideal structure contains a possible relation to spatial extension, every one can be used in thought's comprehension of being. The fact that even the most detached mathematical structure can control reality constitutes the difference between a mathematical structure and every aesthetic system created by the human sciences. Thus we maintain that, like logic, both geometry and arithmetic are ideal sciences.

4. The Science of Thought and Phenomenology

Do logic and mathematics exhaust the sciences of pure thought? One might point out that, in addition to the external being to which mathematics is still related, there is also a being that is viewed from within, a being for which formal laws must likewise be found. This raises the further question about the possibility of a general science of forms, a science that would examine the universal forms of possible objects in complete abstraction from their existence. Such conceptions frequently occur in the history of philosophy and have recently found strong support. *Phenomenology* demands a universal, intuitive comprehension of both the forms of objects and the intentions directed to these forms—a comprehension that will provide a foundation for all other science. It believes that these things can be examined in the same way as logical and mathematical propositions are; it contends that theoretical science can consider the question of the existence of things only when this intuition of essences has been completed.

Doubtless these demands are of the greatest methodological importance, but it is questionable whether they must lead to new sciences, especially to ideal sciences. So far, attempts to establish an ideal science with a value equal to that of mathematics have been unsuccessful; in fact, there are not even any signs of success. This is because psychic realities are completely *qualitative*. Sensations, feelings, and volitions differ from each other in a qualitative way, even when they are quantitatively compared. But relations among qualities do not extend beyond propositions of similarity and equality, and these propositions either say nothing or are conditioned by subjective experience. In the category of quality, thought grasps the element of reality that resists rational presentation, that is existential. Quality is associated with being, just as quantity is with thought. A "mathematics of qualities" as a science of pure thought is therefore impossible.

One can advance the demand for a new science of thought in another way. One can demand the establishment of a "mathematics" of *time* as the form of all internal processes, to accompany the mathematics of space as the form of all external processes. Temporal relations, however, can be established and measured only by means of spatial relations. Time has long ago entered into spatial relations as an element, even as a dimension. A "mathematics" of succession is possible only as an element of a mathematics of extension, because the continuous flow of time is rationally just as indissoluble as are the qualities that are experienced within this flow. Time is associated with being, just as space is associated with thought, and time is accessible to thought only when the existential element time contains is extinguished.

But we must more thoroughly investigate the demand of phenomenology to establish a comprehensive science of thought through the intuition of essences. Undoubtedly the question about the ·existence and explanation of a phenomenon depends upon grasping its *essence*, which appears in an immediate way. This applies to natural objects, and it applies even more to spiritual objects. The reasons modern science has largely neglected this truth are that the foundation of modern science is rational and oriented to the law sciences and that modern science is unconsciously governed by a technical spirit. A thing that has become a part of a machine has thereby surrendered its essence; it continues to exist only as material for a purpose external to itself. A science with this tendency is naturally prone to explain away realities that are not adaptable to this mechanization—for example, life and spirit. The demand of phenomenology therefore symbolizes both an alienation from the technical spirit of science and a reverent return to living reality itself. Indeed, our own placement of the gestalt sciences in the very center of the sciences of being points in the same direction. But it is impossible to establish a new ideal science in this way. For every time it intuits an essence, the phenomenological method is dependent on being. When there are no logical and mathematical structures to be comprehended, there are physical, psychic, or spiritual gestalts to be described in their distinctiveness. But this description is impossible without empirical experience. It is important to observe how the historical, linguistic existence of spiritual objects (as opposed to spiritual essences) plays a crucial role for the phenomenologist, just as external and internal perceptions do with physical and psychic objects. With the exceptions of logical and mathematical structures, the ideal essences of phenomenology are therefore empirical gestalts that are intuited without regard to their existential and causal relations. Thus these essences are infected with the relativities of empirical perception; they do not attain the unconditionality of pure form. Phenomenology is not a new science, but a new spiritual attitude. So the area of the sciences of pure thought is exhausted by logic and mathematics.

Phenomenology belongs to a spiritual situation that is unaware of the *inner infinity* of being in the face of thought. For it, being is merely the material for determination by thought; the only way being resists thought is by preventing a complete realization of thought. There is no resistance, however, leading to new forms and producing the individual and the creative. All positiveness must be situated in form. Such a view was fostered by Greek and medieval thought until Scotism and nominalism appeared. With the advent of these latter movements, this view was transcended and the irrational, dynamic element of being was recognized.

When form is everything, naturally the distinction between the ideal and the empirical sciences loses its significance. Thought is then directed only to itself and to its forms. The external world can at most give stimulus to knowledge; it can never provide material. Knowledge itself is attained by withdrawing from external reality through intuition and the contemplation of essences. Spirit participates in the system of fixed forms, which stand as the essence behind all changing and individual reality and which can be presented in a self-evident, self-contained organization. On the other hand, when the essence is discerned in individuals and the creative process, the system of self-contained forms is destroyed, and the living contradiction between thought and being becomes the principle of the sciences of being. The dynamic view of the world replaces the static view, in both objects and methods.

PART 2

The Sciences of Being
(The Empirical Sciences)

I. Foundation

1. Law, Gestalt, Sequence

In the area of the thought sciences, the *conflict between thought and being* remains latent, for thought is concerned only with itself. To be sure, the sciences of thought establish the forms under which all being must be thought; but these forms do not conceive actual being. Logical axioms express the following situation: the principle of identity merges all being into the pure unity of thought, and the principle of contradiction points to the problem of the "other," the strange, the being that eludes unity. But even this latter principle does no more than point to the problem; the principle remains within the domain of thought; the unity is preserved; the "other" does not become a real problem.

It is different in the sciences of being. Here the "other" is the problem. The conflict between thought and being pervades every empirical cognition. The tension between thought and being sustains the entire system. But the "other" that resists the unity of thought is the multiplicity of *individuals*. Thought desires unity; it creates the universal, the comprehensive, the systematic framework. But being confronts thought as the particular, the incomprehensible, the individual, that which cannot be dissolved in the infinity of thought.

How can the individual resist the universal? Being in itself contains no determinations. Where there are determinations, there is thought. Being would therefore surrender itself completely to thought and its determinations if it could not resist thought. In order for being to resist thought, in order for an individual to distinguish itself from others, being must be filled with thought determinations. The independence of the individual, its power to resist the universal, depends on the degree to which the individual is filled with thought determinations — but (and this is the crucial point) is filled as individual. That is to say, the independence of the individual depends on the degree to which the individual gives all thought determinations its own individual coloration, its peculiar existential character that resists every determination. This is why it is not the unformed being, but the most highly formed being, the *spiritual individual*, that offers the greatest resistance to thought.

From this relation between thought and being, we can now derive three types that determine the material and methodological development of the system of the empirical sciences. In the first basic relationship, thought attempts to confine being completely within its universal forms and thus to extinguish diversity and individuality. We use the concept "*law*" for this relationship between thought and being. We have already referred to this concept in our discussion of the thought sciences. The "propositions" of the sciences of thought are analogous to the "laws" of the empirical sciences. They are similar, because both disregard the individual; they are different, because the thought sciences do not even refer to the individual, but they provide pure forms that are infinitely remote from every individual reality, while the empirical sciences attempt to grasp the individual reality. The propositions of logic and mathematics do not violate things, for they are not even related to them. Physical laws annihilate the individuality of things in order to control them. Law is therefore that goal of knowledge in which the individual is subsumed under the universal.

On the other side, there is a concept that refers to the fact that the individual is inserted into a context, not in order to abolish this individual but in order to represent it. We will call this context a *"sequence"* context. We place the temporal moment within a *sequence*.[1] Even the concept "law" is related to time, insofar as the instances of law are temporal; accordingly, time should not be disregarded when laws are formulated. But in laws, time is, so to speak, only a dimension of space. Time itself has no power to create the new. But time is essentially a category of the new, of development, of history; it is therefore the form of the individual, existential element. The relation between space and time is analogous to that between thought and being. When time is supreme, when it is not merely a dimension of space but is a form of creativity, then being escapes the domination of thought. The sequence context replaces the law context, and time replaces space.

The debate over the methodology and systematics of the sciences has essentially revolved around the antithesis between the concepts "law" and "sequence." The Rickert school stated the terms of the conflict; but the fact that this controversy has not been resolved indicates that the discussion has become sterile and has reached its limit. To begin with, we must ask whether this polarity exhausts the empirical sciences. It does not, for if it did, all sciences would remain permanently restricted. Hardly any empirical sciences would be completely contained by one of the two types; many require a different type. Such a type is derived from the basic relationship between thought and being.

We had assumed that in the sphere of law, being is not completely fulfilled by thought determinations, and that in the sequence sciences, being is completely and individually fulfilled by these determinations. Both assumptions are abstractions, however. Both "law" and "sequence" presuppose "gestalt." The latter concept contains the other two. For every *gestalt* is both an individual and a universal; every gestalt is distinguished from every other gestalt by its individual character and is at the same time the standard for all similar gestalts by virtue of its gestalt laws. The peculiar nature of the gestalt rests on this duality. Law and sequence each realize one of the two aspects that are abstracted from the total gestalt. The law and sequence sciences cognize either universal or particular processes that are not part of a self-contained gestalt. The objects of these sciences are incomplete or open gestalts (a chemical process, a historical series). The contexts they create are, so to speak, linear: they emerge from the infinite and return to the infinite. On the other hand, the gestalt context is circular, so to speak: it represents a self-contained system. Every gestalt is both a law and a member of a sequence series. The more comprehensive a gestalt is, the more it resembles a universal law; the cosmic gestalt would at the same time be the cosmic law. The more concrete a gestalt is, the more it resembles an individual sequence; the absolutely concrete gestalt would be a unique individual within a infinitesimal moment of time. But both the most universal and the most particular are ideas rather than realities.

The problem of gestalt knowledge is easily the most important problem within contemporary systematics. Its solution must overcome the antithesis between individualizing and generalizing conceptual formulations. It can only be solved, however, by considering the objects themselves.

2. Objects and Methods

The contexts of law, gestalt, and sequence are three types of conceptual formulations. But now the question can arise: Are these goals of knowledge valid in the same way for the whole of reality, or are there differences in reality itself corresponding to the differences in the goals of knowledge? According to our presuppositions, there is only one answer to this question: the goals of knowledge correspond to the *forms of being*. The concept "thought" is deprived of its metalogical meaning when it is equated with cognitive reflection. This would be a relapse into logism. "Thought" in the metalogical sense is simply the form of being; all conceptual formulations must correspond to the form of being if they are to be true. We therefore reject subjective idealism in determining the goal of knowledge and maintain that the various groups of being correspond to the various cognitive concepts. When we consider the critical method in philosophy, we will investigate the epistemological problem. But even here we must say that nothing can be reflected upon unless it has previously been thought; that is, nothing can be known unless it has previously been formed by thought, and nothing can be known except according to how it has been formed.

We distinguish three major groups of reality: *physical*, *organic-technical*, and *historical* reality. Within the physical group, there are the mechanical, the dynamic, and the chemical sciences. The second group is divided into two parts, the organic and the technical sciences; the organic group contains biology, psychology, and sociology. Within the historical sciences, we distinguish political history, biography, and the history of culture; this distinction is made less according to the material than to the orientation of the work. Just as the law sciences are related to the propositions of logic and mathematics, so the historical sciences form a transition to the creations that are the object of the human sciences. The area of existential reality lies between propositions and creations.

Therefore, the *physical* group uses the method of law; this group is the domain of spatial contexts. Every law expresses a relation among objects, which are determined only by these relations and can accordingly be exchanged arbitrarily. *Every* mass is related equally to *every* mechanical law, *every* unit of electrical power can be substituted for its quantitative equivalent, and *every* part of an element follows the same laws of repulsion and attraction.

The fact that the physical sciences eliminate everything particular shows their close relation to mathematics — or, speaking in terms of the object, shows that the aspect of being that is observed by these sciences is free of determination.

The *organic-technical* group is completely different. Whether they are organic or technical, the sciences within this group deal with completely formed being. In the organic sciences the gestalt is immanent in things, in the technical sciences it is subjectively posited; in both cases, the object of knowledge is a self-contained context. The individual parts of this context are members that have no reality apart from the whole within which they stand. Their quality derived from the fact that they are members of a whole. Whether it is organic or technical, a gestalt is therefore indivisible. It is possible, of course, to dismantle a machine and to reassemble it, but it is impossible to dismember it while disregarding the relationships of its parts—that would destroy it. Even in dismantled machines, there is an ideal pattern for the relationships among the various parts, although these parts have in fact been spatially separated. A gestalt is therefore a multifaceted and qualitatively determined being whose parts are not quantities, but qualities, or members of a whole. Therefore, a gestalt cannot be created from parts: this statement applies not only to organic gestalts, but to technical ones as well. In the construction of a tool, for example, the important thing is its purpose, not its parts. The purpose determines the formation of every part as a member of the whole. The difference from the organism is merely that in the organism, the formative idea is inherent in the material, but in technology the idea is imposed upon the material. Even this distinction disappears in organic technology, where the purpose is both immanent and transcendent and where the posited goal is only to realize the inner tendencies and possibilities of the organism itself.

Historical reality and its appropriate method of constructing sequence contexts appear when a gestalt contains an element that cannot conform to a law context and yet is completely determined by thought. Spiritual creativity is such a reality. Creative individuality is the object of an investigation in which even structural laws and generic concepts are insufficient. But everything creative is posited in time; it breaks through the simultaneity of space and creates temporal sequence, which is different from physical time. Thus history, the method of constructing sequence contexts, is that investigation of spiritual individuals undertaken by empirical science.

3. Autogenous and Heterogenous Methods

We have just discussed the inner relation of the three basic methods to the three major groups of being. But this treatment does not exhaust the problem of method. We have become acquainted with the native areas of these methods. But the methods are not content to remain within these limits; they transcend these limits and seek to conquer neighboring areas. Every method strives to absolutize itself. *Methods are imperialistic.* This phenomenon is highly interesting for systematics and has been very influential in the history of the sciences. The struggle of methods for predominance has appeared in all areas of knowledge; it has left behind both a profusion of fruitful problems and some devastating effects for the total sphere of knowledge. Obviously, the conflict among methods is a consequence of the basic antithesis between, thought and being: the unity of thought seeks to absorb all being into itself. Can this tendency be justified, does the reason for methodological imperialism lie in the objects themselves? This is indeed the case. The three groups of being are not homogeneous, of course. The first and the third groups are not completely independent; they depend on the isolation of the elements of the second group. This second group is not a synthesis of the other two, as our exposition might have seemed to imply; rather, the two others are dependent on the elements of the second. Only the gestalt, or thought-formed being, exists. To assert the existence of pure being or pure thought is mythology. Laws and sequences can occur only in gestalts.

This relationship among the three groups of being means that gestalts are present in all three areas of being—in the physical group, of course, gestalts that are not yet self-contained, in the historical group, those that are no longer self-contained. But where there are gestalts, there are also laws and sequences, so that elements of the two other areas can be found in each of the three areas. This justifies the striving of every method toward universalization.

But the limits of the methods remain insurmountable. It is impossible to understand in a concrete way physical or historical events as the expression of a total gestalt. Reality as a total gestalt is an idea. And it is impossible to treat the structural tendencies found in physical reality as complete gestalts or to treat structural consequences in history as genuine gestalts. Naturally, then, one cannot adequately apply the law method to history, or the historical method to physics.

We express this dual relationship between methods and objects by speaking of the "*autogenous*" and the "*heterogenous*" application of methods. Methods are autogenous within their native area; here they are adequate to their objects. But they are heterogenous when they encroach on foreign areas, where they are adequate to only one element of the object, not to the object as a whole. Both the wealth of the system of the sciences and the necessity for systematics continually to redraw the boundaries are based upon this distinction between autogenous and heterogenous methods.

In the physical sphere, the tendency toward gestalt appears in two ways: first of all, the view of reality as a whole is conditioned by the idea of a *universal gestalt* to which all particular processes are organically related; in the second place, the physical processes themselves are inconceivable without *structural elements*, and these elements assume an almost organic character in organic chemistry and crystallography. Where there are gestalts, however, there are also sequences. Thus the idea of a macrocosmic gestalt corresponds to the *history of the universe* and of its parts, a history that forms a transition to geography, proceeding through astronomy and geology. And corresponding to the microcosmic gestalts, there is a history of the appearance and disappearance of the elements, a *history of energy and matter*. But this history has not yet become an independent science.

In the physical group, the autogenous law method is fundamental. This group also contains the heterogenous methods of both the theory of macrocosmic and microcosmic gestalts and the history of celestial bodies and matter. Because these gestalts are imperfect and incomplete, the gestalt method has not produced an independent science here; by comparison, the history of celestial bodies is the object of a variety of sciences. Macrocosmic and microcosmic gestalts are examined within the context of laws and sequences, because they are never the object of special consideration. For the gestalt element is always presupposed, and our detailed exposition will show how this element becomes increasingly evident in the sciences that form a transition to the organic group.

We divide the physical sciences into two types: the generalizing sciences (mechanics, dynamics, chemistry) and the individualizing ones (astronomy, geology, geography). In both types, however, we indicate the structural elements that they contain and that constitute the essential methodological differences among the individual sciences.

In the organic group, the gestalt method is autogenous and completely dominant. The other methods have not developed their own sciences, but they operate within the individual sciences of this group. Nevertheless, we must make an important distinction: the establishment of gestalt laws is not an invasion by the pure method of law, nor is the establishment of individual species concepts an invasion by the pure sequence method. The gestalt method unifies the law and sequence methods. The law method makes its first heterogenous appearance when the development of gestalts is *explained* by means of universal physical laws, when laws are primary and gestalts secondary. And the heterogenous influences of history appear both when the *theory of development* becomes the history of biological, psychological, and sociological forms and in the history of technology.

Finally, in the historical group, the heterogenous methods lead to two distinct ways of examining historical reality: the *doctrine of spiritual gestalts* and the investigation of *historical laws*. Once again we face the question of the entire structure of reality, the question of whether law or sequence is ultimate. But this question leads beyond history to metaphysics.

4. The Dispute over Methods

Now that we have established the systematics of the empirical sciences, a glance at the status of the debate will show that our conception can solve the actual problems. Currently a methodological and an objective position are locked in combat. The methodological position of epistemological idealism divides the sciences into *the natural and the cultural sciences*. The objective position of epistemological realism divides them into *the natural and the human sciences*. For the first position, psychology belongs in the natural sciences, because its methodological procedure of generalization is similar to them. But from the perspective of the second position, psychology is the foundation of the human sciences, because it deals with the same object as they do, the life of spirit. The placement of psychology is thus the criterion for defining both positions. Therefore, this apparently formalistic dispute has a concrete significance: it determines the fate of the human sciences, the conception of spirit and culture. If psychology is considered the foundation of the human sciences, then spirit is no longer the realm of the individual and unique, creative sequence becomes structural law, thought destroys being, and rational form triumphs over the irrational import that contradicts it. The methodological position avoids this mistake, but it itself suffers from numerous deficiencies. It does not distinguish between history as an empirical science, and the pure, systematic human sciences; it forces the latter likewise toward a rationalistic conception that overlooks the creative nature of spirit. Moreover, this position does not answer the objection raised by the objective method, that psychology is not a physical science. It cannot meet this objection, because it overlooks the central area composed of the gestalt sciences, which contain psychology. Finally, it cannot do justice to the historical elements in the physical and organic groups, because it limits the historical method to the cultural sciences and does not recognize the distinction between the autogenous and heterogenous methods. Reality is too rich to be divided between two methods; the gestalt method, which was forgotten in the methodological dispute, is really the central, concrete method, the method that is appropriate to thought-formed reality and that is thus the only one able to solve the methodological problem.

We have excluded the concept "*natural science*" from the systematic treatment of the sciences. Of course, nothing prevents one from uniting the series of sciences running, say, from mathematics to biology under this name; but that would be a practical classification, not a systematic one. From the systematic point of view, mathematics belongs to the thought sciences, biology to the gestalt sciences. The concept "natural science" (especially in contrast to the human sciences) has caused so much methodological harm that it is time either to replace it with "science of being" or to restrict it to the physical group while excluding mathematics, biology, and psychology. For the mathematical sphere is not yet nature, and life and the psyche are no longer merely nature.

5. The Attitude and Procedure of Knowledge in the Empirical Sciences

When we discussed method in the last section, we were referring to the goal of knowledge, not to the other sides of the
methodological problem we considered in the "Foundation" of
the thought sciences. Now we turn our attention to these other
sides.

The basic attitude of all empirical sciences is the submission
of knowledge to being. In the empirical sciences, the object
confronting knowledge is strange to it. Knowledge can attain
its goal, not by attending to its own inherent forms, as it does in
the thought sciences, but by leaving itself and directing itself to
the "other" that stands at an infinite distance. Instead of intuition, here there is *perception*. In intuition subject and object
are one, in perception they are two; they must indeed become
one in the act of knowledge, but they can never attain perfect
union. Every perception is an absorption of the object by the
subject, but an absorption that always remains finite and
limited in comparison with the inner infinity of every existing
gestalt.

Perception becomes knowledge when reason establishes contexts. We call a rational perception "*experience*," and according to this basic attitude, we could also call the empirical
sciences "the sciences of experience." That which is rational intuition within the thought sciences is rational perception, or
experience, in the empirical sciences. The duality of thought
and being is reflected in both activities; formative activity corresponds to thought, and the receptive absorption of material
corresponds to being, whether being is ideal or real.

But there is still a third element influencing the attitude of
empirical science, an element that is derived from the sphere of
spirit and that has no place in the thought sciences. In order to
characterize it, we must borrow two terms from the
methodology of the human sciences. In the human sciences,
thought determines itself as fulfilled by being. Thus something
that is absent in the empirical sciences occurs here: subject and

object are united. But the situation in the human sciences differs from that in the thought sciences. Though in the latter the objects of knowledge are the mere forms of thought, in the former the objects are the spiritual forms that are filled with being, forms that contain the infinite and irrational of everything in existence. A cognitive attitude that achieves union with such objects we call *"empathy"* (speaking in terms of the emotions) or *"understanding"* (speaking in terms of the intellect). In the human sciences, understanding is self-creative; in the empirical sciences, it is re-creative and perceptive.

These three elements are divided in different ways in the three groups of the empirical sciences. In the physical group, the rational element is most prominent; in the historical group, empathy is strongest; but in all three of them, perception is fundamental and essential.

This relationship becomes even clearer when we consider the procedure of knowledge, which corresponds exactly to the methodological attitude. Perception becomes knowledge through *description*. Description is the foundation of methodology in all the empirical sciences. The great service of phenomenology was that it emphasized description as the presupposition of all explanation and construction. Whether they are incomplete or complete, gestalts are objects of description. From the abundance of perceptions, description extracts those elements that constitute the gestalt as a distinct one, those elements on which the gestalt context depends. The individual element of every gestalt, the element that is also the common feature of all similar gestalts, is described.

Empirical description is analogous to demonstration in the thought sciences. Instead of the deduction found in the sciences of thought, however, in the empirical sciences there is a peculiar union of deduction and induction called *explanation*, which finds its most perfect expression in the methodical investigation of being, that is, in experiment. Just as perception is unfruitful unless it becomes experience, so is description unless it leads to the comprehension of context, and so is induction unless it is combined with deduction to become explanation.

The third element derived from the human sciences is *construction*. Although in the human sciences it, as creative construction, creates normative systems, in the empirical sciences it is re-creative, descriptive construction.

Obviously, the method of explanation is most important in the physical sphere, where it is tempted to become pure deduction, without of course being able to do so, as long as it seeks to know existents. On the other hand, the method of constructive re-creation is essential to the historical sciences, where it is tempted to become normative construction, thus relinquishing its empirical character. Both methods are always based on description, however; description grounds and balances both methods within the organic-technical group. Explanation and construction are description.

Finally, we must answer the question of the degree of knowledge within the empirical sciences. What corresponds to the self-evidence of the thought sciences? The answer can only be *probability*. Probability has degrees, in accordance with the relationship of infinite approximation between all knowledge and being. But how can there be any degree of certainty if there is an infinite gap between knowledge and being? Mere perception could not yield even probability if it did not become experience through reason. The deductive element of the empirical method yields the degrees of probability. The fact that existents correspond to the forms of thought makes empirical science possible, and the degree of probability in knowledge corresponds to the degree of strength possessed by the deductive element. In mathematical physics, probability approaches self-evidence without being able to attain it.

Here, too, we find a concept from the human sciences, the concept "*conviction*." This concept expresses the kind of certainty possessed by the constructively creative attitude. Conviction is crucial for all the normative sciences. In the empirical sciences, it finds pure expression as little as does the concept "self-evidence," but it is strongly influential in these sciences (e.g., in the historical group). The peculiar mixture of probability and conviction is characteristic of the historical sciences. Both self-evidence and conviction are combined in the central group of the empirical sciences, the organic-technical group.

This is the methodology of the empirical sciences. At every point, this discussion shows both the independence of the three groups and their mutual dependence, which rests upon the fundamental significance of the gestalt group.

6. Categories and Methods

The philosophy of knowledge, that is, the theory of the nature and categories of science, shows that objects and methods have a common foundation in the *categories*. Categories are the spiritual functions constituting the world of phenomena. The possibility of knowledge depends on the categorial connection between being and consciousness. The principle "Only that which is thought can be reflected upon" is epistemologically grounded in the theory of categories.[2] The fundamental category of the empirical sciences is that of *causality*. Relationships between existents are causal relationships. All other categories, including space and time, can be understood from the perspective of causality. But causality itself expresses the original relationship between thought and being within the sphere of being.

The logistic view of causality extracted the purely formal definition of causality, as a necessary sequence in time according to a rule, from the relation between cause and effect. Thus it stood in legitimate protest against empiricism's conception of causality as a subjective habit developed in response to frequently recurring consequences; for the concept "habit" already presupposes causality, just as does every other concept that explains causality. The logistic position was also correct when it rejected every notion of a mythological power that was supposedly causative. But it overlooked the fact—and necessarily so, from the logistic standpoint—that the mythology of power did not intend to read physiological feelings of tension into things. The real intention of this mythology was to point to the irrational element that is grounded in the existential nature of things and that is contained, with the rational element, in the causal relationship. We can view this metalogical element of causality from two perspectives: first, as the selfhood of every individual existent over against every other existent, as the resistance it offers to any assimilation; second, as the common existential root that is revealed in the system of effects. Time does not separate causal connections from ideal structures; "being" does.

The logistic conception is naturally most appropriate within the sphere of mathematical physics. Here a formal, abstract definition is sufficient, because the sphere itself is abstract. In the physical sphere, causality is *quantitative* and subject to the law of *equivalence*: the effect contains nothing that was not present in the cause. Thus, the law is abstracted from the qualitative particularity in the reactions of individual things; it only attends to what is quantitatively measurable. Causality in the historical sphere stands in sharp contrast to this. Here there is no equivalence of cause and effect at all. *Quality* and the law of *production* prevail: there is something new in the effect that was not present in the cause; the relation between cause and effect is not a rule, but a sequence of meaning. In the face of this fact, the logistic definition utterly collapses. Existential independence and relatedness are revealed as productive causality. In place of equivalence and necessity, there is individual creativity, or freedom.

Equivalent causality represents a relationship of quantitative exchange. The abstract objectification of this relationship is "*matter*." Matter is the objectified substrate of the quantitative relations, a substrate with all qualities excluded. In this form, it symbolized substance. But the intention of the category of substance, to say that all real things are rooted in being, was itself made into a thing in this way. As in the case of causality, the critical reaction to this false objectification has led to a logistic, formal treatment of *substance*, a treatment that overlooks its metalogical meaning. It is impossible to make matter symbolize substance. That which supports things is certainly not that which remains after the subtraction of all qualities for the purpose of making quantitative relationships of exchange possible. That which supports things and makes equivalent exchange possible is the existentially rooted, living form of things. Matter is an abstraction from these realities; it is not the reality within them.

The category of substance should be used when causality presents a self-contained relationship. "Substance" expresses the unity of a self-contained causality, a unity that is the basis of everything individual. But self-contained causality is the presupposition of equivalent as well as productive causality; it contains both elements. The category of *purpose* has been used for self-contained causality. Here, too, we must avoid a false mythology that objectifies purpose and makes it a kind of mechanical causality. The concept "purpose" in the metalogical sense means the independent, self-contained character possessed by causality in the gestalt sphere. For every gestalt reality is a unity of equivalent and productive causality. It is productive insofar as it posits an independent, qualitative gestalt; it is equivalent insofar as the quantitative exchange occurs within this gestalt. All causal series are based upon substances; that is, all causal series are ultimately teleological.

But this ultimate teleology is an idea, not a reality. The open character of the equivalent and creative causal series prevents it from achieving reality. This leads to the heterogenous use of the category of substance for both matter and spirit. Matter symbolizes the open causality of the physical sphere; spirit as substance symbolizes the open causality of the historical series. But both are substances in the heterogenous sense; the self-contained causality of the gestalt reality is substance in the autogenous sense.[3]

Thus the consideration of the categories is related to that of methods, providing the ultimate foundation for the division of the empirical sciences into the groups of law, gestalt, and sequence.

II. The System of the Empirical Sciences

A. First Group: The Law Sciences

1. The Autogenous Series of the Physical Sciences

a. Foundation: Mathematical Physics

Mathematical physics forms the proper transition from the thought sciences to the sciences of being. For centuries it has been viewed as the very model of science. It is the most grandiose attempt undertaken by thought to master being; in this science, thought strives to grasp being completely in mathematical forms. The degree of its success marks the victory of rationality over reality.

Of course, it had to be clear from the beginning that what mathematical physics grasps is not perceptible reality, but an abstraction that is detached from all perceptible qualities: extended mass moving in space. This science regards colors, sounds, and the abundance of perceptible reality as secondary, subjective qualities. It assumes that this abstract reality is the true reality. But to us it is clear that mathematical physics treats only a certain element of being: the quantitative relations of equivalent causality.

The *abstract* nature of mathematical physics follows from this. The science is not mathematics, for unlike mathematics, its objects are actual bodies moving in space. It is not a science of thought; it is an empirical science, and it can resist every attempt to consider it a mere thought science by showing that it can predict and control reality by means of its calculations. But it is an abstract empirical science: it singles out one element of being and disregards everything else. Nevertheless, it is not heterogenous, for in the physical sphere the quantitative relations of exchange are essential. Idealists are therefore mistaken when they deny that mathematical physics has any empirical knowledge. But their opponents are just as mistaken when they claim that mathematical physics can grasp individual reality. Every part of reality is qualitatively different from every other part. One cannot deprive even the atom of all qualities; a kind of individual structure must also be attributed to it. But a mathematical formula cannot take account of an individual atom. Even when expressing a physical law, the formula is infinitely distant from every concrete thing and process. Of course, the formula can apply its abstraction to every part of reality; it can calculate every individual. It does not approach the individual insofar as it is an individual, however, but only insofar as this individual is a special instance, quantitatively determined and without individuality, of a universal law. Even with this restriction, knowledge of the individual is not born from an interest in the physical world. This knowledge either has an experimental aim and is then the means, not the goal, of knowledge; or it has a technical and descriptive intention, that is, it is inserted into a gestalt context or sequence context. Physics itself has no interest in the individual.

Since mathematical physics is a science of being, it has an empirical, inductive character, not the self-evidence possessed by mathematical structures. But it is presented in a mathematical, and thus self-evident, form. This apparent contradiction reveals the nature of this science even more clearly. That is to say, it shows that in its quantitative relations, reality is in fact mathematical. The quantitative relations are abstractions from the whole of concrete reality, but they are not just abstractions; they correspond to reality. Based as it is on the fundamental relation between thought and being, this fact has allowed mathematical physics to become the *ideal of pure science* as such. Thought was ecstatic about its initial successes. Here it had acquired a method that realized all the fantastic dreams of the early Renaissance and that strove to exploit reason's luck with mathematics for a conscious victory over reality in general. No object was to remain unsubdued. The certainty previously provided by the authority of the sacred was now to be provided by the intrinsic self-evidence of thought itself; this self-evidence was to be found in mathematics. The method of mathematics, its uniformity and power of conviction, was to be valid for all reality. Thus all areas of being were in succession claimed for mathematical physics: the living was killed so that it might fit into mechanical formulas. The psychic dimension was either reduced to mechanical elements or made into a process accompanying material mechanics. A statics and dynamics of the emotions attempted to apply mathematics even to the life of feeling. Spiritual values were treated according to a technological model; they were constructed from the simplest elements. Metaphysics itself deduced the world *more geometrico*.

Much philosophical discussion has been occupied with the conflict over mathematical method in the empirical sciences, a conflict that has become the source of countless problems. The distinction between autogenous and heterogenous methods explains and resolves this conflict. And this resolution accords with the basic relation between thought and being.

b. Mechanics and Dynamics

Mathematical physics celebrated its first great triumph in *mechanics*. For here it dealt with an object (homogeneous, moving mass) that is formed, like mathematical space, by definition. From this fact some have concluded that mechanics is not really an empirical science: it is an ideal science belonging, with the theory of motion, within mathematics. But this view is unjustified, for mass is rooted in being, pure *motion* is not. Mass is neither a structure nor a function of a structure; it is an abstraction from reality. Its abstract nature does not disprove its existential reality. However completely mass may be subject to mathematics by means of definition, it cannot be derived from mathematical presuppositions. The empirical element in mass is indissoluble; this element is expressed in all the laws of mechanics.

Mechanics is not concerned with the effective forces that move mass; the sciences of dynamics are. Though it is a mathematical science, *dynamics* is an area in which the qualitative element can no longer be disregarded. Of course, even here the mathematical element has the broadest latitude; all dynamic relations can be expressed in quantitative terms, and mathematics has increasingly succeeded both in eliminating the peculiar qualities of the individual physical areas and in comprehending all the forces and materials as appearances of a single force. Every new insight has brought new victories to quantity and has strengthened the rationalization of existents. Still, the areas of dynamics add something new to the pure formulation of law. The real forces of mass (e.g., light, heat, electricity) cannot be produced by definition, as can mechanical mass. Rather, they are existentially given; they cannot be completely dissolved by thought. Consequently, dynamics is not without qualitative and descriptive elements. Yet even if it wishes to reduce the qualities of its forces completely to quantities, dynamics proceeds from these qualities and can finally eliminate them only by inserting them into the subject. But even with this, dynamics has not said its last word. Because it aims to know realities instead of possibilities, it cannot abstract from one ultimate: the structure of dynamic substance itself.

Quality is not merely a subjective category. Every statement about velocities, polarities, and reciprocal actions of dynamic substance expresses a qualitative proposition within a quantitative formula. The existential structure of forces does not necessarily conform to thought; the givenness of matter is present in this structure. Of course, this structure lacks an inner teleological impulse. The total structure of reality remains hidden. Therefore, one can only describe structural elements; one cannot represent a structure as such — every attempt to do so is fantasy and mythology. Dynamic physics is a law science. Qualities are considered only at a preliminary stage. The gestalt method remains heterogenous.

c. Chemistry and Mineralogy

The problem of gestalts leads to *chemistry*. Just as dynamics deals with forces, so chemistry is concerned with matter. But this contrast is only apparent. The question of the gestalt of the ultimate particle of matter, the atom, leads directly into the area of dynamics. The analysis of the atom into electrical substances abolishes the distinction between force and matter, placing chemistry within the distant region of dynamic physics. One might separate this area, as the theory of microcosmic gestalts, from both mathematical dynamics and descriptive chemistry. But the impossibility of directly perceiving the smallest and largest gestalts, the necessity of relying on the conclusions drawn from observed relations, the inner infinity of the task itself — all these factors make such a separation impractical. The infinity of being in the face of thought becomes clear in the infinitely small and the infinitely large. The infinite breaks through every gestalt, however. Thus the *an-*

tinomy of gestalt, an antinomy that is based upon the fact that we must think of reality in terms of structure and yet cannot do so because of its inner infinity, appears as the foundation of the Kantian antinomies of space and time.

The problem of the microcosmic gestalt also leads necessarily to the problems of sequence. Modern chemistry has increasingly abandoned the notion of fixed elements; under the influence of radio dynamics, it has focused on the emergence and dissolution of the elements. It thereby encounters a problem already apparent in dynamic physics, especially in the theory of heat: the problem of the loss of heat, or entropy. It is impossible to deal with these questions merely by using law concepts. Either one must presuppose that the universe has had a definite original state, or one must consider the universe as perpetually creative. However far back one pushes the moment of the mere givenness of pure being, somehow this moment appears in every instant and continually raises the question of a productive sequence series and suggests the idea of a *history of force and matter*.

Dynamic chemistry has proceeded from *descriptive chemistry*, in which the problem of quality is much more significant. However far the analysis of matter goes, the synthesis will finally become possible and the abundance of quality will be explained. The elements are distinguished by their properties; they have affinities and peculiarities; they enter into combinations that create something qualitatively new. Chemical formulas disclose the gestalts of phenomenal material. They symbolize the omnipresent elements of a cosmic gestalt. Nevertheless, these formulas do not reveal a total gestalt, for they are isolated from each other and are subject to the regularity that ignores individuality. They belong to the physical sphere, in spite of the heterogenous influence of the gestalt method. With regard to alchemy and the idealist philosophy of nature, it is important to note that the gestalt method is heterogenous in chemistry. Otherwise, a fantastic view of nature is inevitable.

Organic chemistry stands nearer to the genuine gestalt method. It forms the transition to biology. But organic chemistry itself is not biology; it is physics. For it abstracts from life, the material of which it analyzes. And it does not even approach the gestalt when, as *synthetic chemistry*, it shows how to produce organic matter. Even the matter of organic chemistry is subject to the general concept of law. Yet it is interesting to see how the structural elements of organic substances exhibit an extremely complicated structure, without being able by themselves to lead to a genuine gestalt. The organic gestalt is the prius of organic matter, not conversely.

The theory of crystallization is a member of the chemical group, though it occupies a special place. The *crystal* belongs to the physical sphere, but it is distinctive because it represents a structure that is concretely perceptible. The crystal is indivisible and uncompounded. It is grounded in the microstructure of dynamic substance. Nevertheless, the gestalt character of the crystal is heterogenous. The crystal lacks the inner teleology, the organic connection of the individual parts, and the individual character, of genuine structure. Therefore, the part of a crystal does not lose its gestalt character when it is separated from the whole, as the part of a genuine gestalt does; it remains a crystal. The gestalt is not constitutive. It is not substance.

Whereas organic chemistry and crystallography lead into the organic group, *mineralogy* forms the transition to the heterogenous series of the physical sciences. Mineralogy is the description of those compositions of matter found on the earth. Thus it has both generalizing and individualizing aspects. It investigates the laws of existing compositions of matter and is thus an extension of chemistry; it describes this matter from the point of view of its geological and geographical reality. Thus it leads to these sciences themselves.

2. The Heterogenous Series of the Physical Sciences

a. Foundation: The Individual Element in the Physical Group

We have shown that the gestalts in the physical group are not genuine gestalts, because their parts are not determined by the whole and thus they lack organic connection. Accordingly, there can be no genuine sequences in this sphere, because the particular has no inner *individuality*. Here the individual is determined by position and quantity; it does not create anything original. The sequences into which the individual is inserted are external to it; they do not establish its essence, as they do in genuine sequence frameworks. But how are individualities demarcated here, and from what viewpoint are sequence contexts created? These questions are both important and difficult.

The systematics of idealism answers that a thing can be considered an individual when it is related to *culture* and thus to genuine sequence contexts. There is certainly a connection between the geographical object and the life of culture, for example. But this connection neither determines the research interest nor influences the selection. In geology, this connection is much less significant. Formations of the earth's crust, which are seldom even indirectly related to culture, stimulate the same scientific interest as the formations of cultivated lands do. The moon and the stars are not related to culture at all. Consequently, an objective systematics tries to make *magnitude* the principle for selecting individuals. But magnitude is a relative concept, and one must always ask: Large in comparison to what?

The real principle of selection is the constitutive significance of a thing for a *context*. The object of the heterogenous science of sequence is whatever is indispensable for completing a context. This explains how individuals are distinguished from each other in the general area of the physical, despite the fact that the principle of individualization is not inherent in the things themselves. It is the context that delineates the individual. Whatever manifests itself as independent within the context is thereby individualized: for example, the constellations are significant within the cosmic sequence, but the atmospheric strata or the mountains are irrelevant for this context. On the other hand, even the smallest parts of mountains are important in geography when these parts (e.g., a road, a group of trees) appear within the context of a particular landscape.

But what kind of context should be selected? To be sure, one could assign geography the task of ascertaining the structure of the surface of all stones and leaves, as well as the structure of every atom, the number and position of all electrons in every moment — that is, of comprehending the uniqueness of every moment of time in every relationship. Philosophy has often claimed that this ideal is realized by an intuitive understanding that is conscious of both the universal and everything particular. But the human spirit does not have this intuitive faculty. In order to contemplate the one, it must abandon the other, which meanwhile changes and remains unknown; and it cannot perceive the alterations within the domain of the infinitely small. Here we encounter the limit of the human *cognitive standpoint* — a limit, incidentally, that would apply to every possible standpoint.

In addition to this fundamental, insurmountable limit of individual knowledge, there is a concrete, practical limitation: man is not only a knowing being, he is also a living being, and his will to knowledge is limited by his will to live. Only that which can somehow enter into the context of human life, whether through use, through emotional influence, or through relations to spiritual values, actually becomes an object of knowledge. This is the *pragmatic* nature of human knowledge. The pragmatic element is not the formulation of concepts, as philosophical pragmatism contends, but the selection of objects. This view avoids the onesidedness of both the systematics based upon methods and the one based upon objects. The relation, not only to spiritual values but also to every conceivable kind of life values, selects the sequence contexts. What is crucial for the direction of knowledge is not an abstract magnitude, but the magnitude that establishes the relation to the knowing subject as a living being. This form of pragmatism is superior to both other theories because it provides a definable criterion that is both comprehensive and infinitely variable, yet always certain.

b. Cosmic Gestalts and Sequences

The infinitely large and the infinitely small gestalts cannot be known directly. But we are indeed confronted with cosmic gestalts that are also individual, namely, stars. The knowledge of stars initiated a special science, *astronomy*. Astronomy is a heterogenous gestalt science, in two senses. On the one hand, it investigates the gestalt relations between the stars and thus resembles a theory of the macrocosmic gestalt — a theory that is unattainable. On the other hand, it examines the gestalt of constellations themselves, attempting to establish the structural laws of their composition and evolution. The first form of investigation is indissolubly linked with mechanical and dynamic physics, being their principle application; the second form is the most important heterogenous use of the gestalt method.

Constellations are undoubtedly closer to being gestalts than is anything else in the physical sphere. This is so obvious that there have been repeated attempts to conceive constellations as genuine gestalts, in the biological and psychological senses. But even these attempts clearly indicate the heterogenous character of the cosmic structures. Like crystals, constellations lack the immanent teleology by which isolated parts become members of a whole, members that lose their essential nature as soon as they are separated. Constellations can be destroyed, but the parts that are destroyed follow the same cosmic laws as those that are not; all the parts are celestial bodies.

In the field of dynamic chemistry, the investigation of microcosmic structures has led to the idea of a history of energy and matter. But this idea can only be realized in astronomy, where individual gestalts invite a consideration of the whole structure of reality. Constellations are gestalts that evolve along the same general lines, but in individual ways. The universal law of evolution leaves room for characteristic peculiarities, both in these and in genuine gestalts. This is why one can establish various sequence series, both in the evolution of the individual constellations and for the entire celestial structure. The history of energy and matter is realized in a *history of the heavens* and of its structural elements, the constellations.

c. Geology and Geography

Cosmic history leads to *geology*, the history of the earth's evolution. Geology is continued in *geography*, the description of the earth's surface. As *paleontology*, geology investigates organic remains of the past; geography, as the geography of plants and animals and as anthropological geography, encroaches on the area of organic-technical reality. Thus the methods of these two sciences overstep the boundaries of the physical sciences. Geography nevertheless belongs within the physical group. The science of law embraces all living

creatures; its laws are valid for falling stones as well as for fly-
ing birds. But here there is no reason to give special attention
to these objects; the individualizing series deals with these
variations. Organic-technical things are not treated in an
organic-technical way, but in a geographical way, from the
point of view of their diffusion, their existence, and their position on the earth's surface along with the types of rocks,
oceans, and deserts. Thus the method of geography extends
beyond the physical sphere into sociology and history, where it
enters the genuine sequence science of history.

Geography is the description of the earth. To describe
something from the geographical point of view means to conceive it as an essential gestalt element of the earth. But the
geographical description of the earth is defined by a concept
peculiar to it, the concept "*surface*." "Surface" in the
geographical sense is not merely an empirical concept; though
it is not a genuine category, it is similar to a category.
"Surface" is a pragmatic category, formed by proceeding from
the relation of life to things. To the degree that thought forms
being, things receive an *interior* and an *exterior* — indeed, the
genuine gestalt is directly determined by this distinction. Thus
the closer a creature comes to being a genuine gestalt, the more
it contains this contrast. The contrast is most distinct in the
relation between psychic inwardness and corporeal outward-
ness. But the closer a creature comes to this relation, the more
the exterior expresses the interior, and the less the creature has
its own special significance. Consequently, apart from its
character as the expression of the interior, it provokes no scientific interest; it is thus excluded by the pragmatic principle of
selection from any special treatment by geography. It is different with the gestalts of the physical sciences; they are not genuine. These sciences have a pragmatic interest in the surface,
for their gestalts are pragmatically related to the knowing subject through their surface. Here the contrast between interior
and exterior is not qualitative, for every part of the interior can
become a surface when the outer layers are removed. But
everything that has a surface and is thus significant for the formation of the earth's surface belongs to the surface of the
earth: plants, animals, peoples, races, technology, political institutions — all of these have an aspect that is geographically

relevant. They help condition the gestalt that is the surface of the earth. This is the universal significance of the geographical method, though it is often overlooked.

When it does not extend into dynamic physics and chemistry, the method of geography is descriptive. The method of *description* is especially suitable for clarifying the nature of description: to describe is to create a gestalt context. In description, those elements that condition the context and are essential to it are selected from mere appearance. There is no description without the idea of a gestalt, and when there is a gestalt, there is description. But the surface gestalt is distinguished from the interior gestalt by the fact that the former is not subject to structural law. When geography formulates laws, it becomes astronomy, geology, or physics. On the other hand, pure geography has no laws. But description is impossible without a rational element. This element is not one of law, it is one of construction. And so instead of gestalt law, there is *gestalt construction*. Every geographical description aims at a gestalt construction — not a construction proceeding from inside, which would be re-creative, but one from outside, which is pragmatic, determined by the living connection with the knowing subject. Understanding is also a part of the geographical method: there is even a geographical empathy, but it is not genuine empathy, for it does not empathetically project itself into the gestalt of being, but into the subject-object relationship. This corresponds to the fact that the sequence and gestalt methods are heterogenous when they are applied to geography and its entire group.

B. Second Group: The Gestalt Sciences

1. The Organic Sciences

a. Foundation

The distance from the mechanical-physical group to biology is much greater than that from *biology to psychology*. This contention is not beyond dispute, of course; it requires thorough substantiation. One might claim that this contention would abolish the distinction between *nature and spirit*, a distinction even more decisive than that between stone and plant. And this distinction would correspond to the methodological one: on the one hand there is an external perception, on the other an internal one, so that thought has a fundamentally different attitude toward its objects in the two cases. But there are compelling arguments against this objection. First of all, as far as the biological problem is concerned, one simply cannot make "life," in the biological sense, a pure object of external perception, as one can physical things. It is certainly possible to formulate the essence of life by distinguishing it from physical matter, but this essence cannot really be grasped in this way. One cannot proceed from the physical and objective to the organic and living, though one can grasp the inorganic from the side of life by objectifying the quantitative elements of the organic. Life requires an act of self-comprehension, of understanding.

It is possible to take one more step. The question of the nature of "vitality," and thus of the principle through which the organic is formed, can be answered by attempting to understand vitality according to the analogy of the psychic dimension. This conception prevents the psychic dimension from suddenly appearing within the organic series and immediately having important effects. The psychic and biological dimensions are essentially related: the psychic is the interior aspect, the biological the exterior aspect, of the living gestalt.

The relation between biology and psychology is expressed in the methodological concept common to both, that of "*gestalt*"[4]: the soul, in psychology, is the teleological center of all psychic functions, just as life is the center of all the organic functions. In both cases, there is a gestalt context—not a law for random, unrelated objects, but a context of separate functions for which this context is also their law. It is just as impossible to have a thought or an emotion existing outside a psychological gestalt as it is to have an arm or a nerve outside a biological gestalt. But this phenomenon is so significant and, in contrast to law, so original that it requires a union of all the sciences under its rubric.

One important consequence of our systematics is that it permits us to classify a third science, *sociology*, which otherwise cannot be categorized. Though biology and psychology are related to each other as the external aspect to the internal, sociology is related to the other two as the communal aspect to the individual. Therefore sociology contains biological and psychological elements. But it looks beyond these elements to its own object, the social organism, which includes both biological and psychological elements. Thus sociology is also included in the gestalt sciences. The social organism is a gestalt context. The function of its individual members has reality only within this context. The social organism is a genuine gestalt. The science of gestalt is basically a science of the unique. It considers the *living gestalt*, first from the perspective of its individual exterior, then from that of its individual interior, and finally from that of its social exterior and interior.

The dispute over the classification of the sciences would have been more fruitful if this group mediating between the law and sequence sciences had been acknowledged. Above all, the difficulties involved in the systematic classification of psychology and sociology would have been surmounted. But this union is possible only because the concepts "natural science" and "human science" are excluded from the group of the empirical sciences in general. This exclusion expresses a rejection of Cartesian *dualism*, which is based upon the despiritualization of mechanized nature; it makes life and creativity, the gestalt of being, the prius of all knowledge of reality.

b. The System of the Organic Sciences

i. Biology

The Autogenous Method of Biology. Biology is the arena of a conflict between two fundamental views that have been in contention throughout the entire history of philosophy: the *mechanistic* and the *vitalistic* interpretations of life. On the one side, there is an attempt to dissolve the peculiarity of life by using mechanical-dynamic categories; on the other, there is both a recognition that this attempt is impossible and a contention that there are special vitalities governing the biological process. This conflict between mechanism and vitalism expresses the conflict within biology between the law method and the gestalt method. Vitalism can appeal to the actual formulation of biological concepts. Almost all biological propositions express, directly or indirectly, the teleological relation of all organs and functions to the unity of life. On the other hand, mechanists can appeal to the fact that teleology is never a substitute for causality, in the case of individuals.

Purpose neither is nor should be an explanatory principle. Yet purpose is the principle that completely governs biological statements. Organs and functions of life cannot be defined without reference to purpose. They are what they are only in the living unity that connects them. The strictest advocates of the quantitative-causal method never speak of causal series that emerge from the infinite and return to the infinite, accidentally intersecting the living organism; they look, rather, for causalities within the living gestalt. And the proponents of the vitalistic method try to solve the gestalt problem by the concept "life principle." Of course, the vitalists thereby usually commit the error of making the life principle an equivalent cause. Then the life principle becomes the cause of formation, or shaping. In this way vitalism bases itself on mechanism, admitting that the latter is the primary standpoint. This is justified only insofar as vitalism expresses the resistance of existential life to mechanistic rationalization, but not insofar as it introduces a new, mystical principle of causality. The gestalt is the prius of gestalt laws. This is the only truth of vitalism.

In biology, the establishment of universal *structural forms* through generic concepts stands alongside the investigation of *structural laws*. In the sphere of dynamics, we find spurious generic concepts; the species of the dynamic, the chemical groups and their compounds—these are universal concepts under which every individual object within these areas is subsumed. But the difference from genuine species is that here the individual object has no independence whatever—that, for example, the general concept "oxygen" is not formed by abstraction from the individual attributes of particular quantities of oxygen, but that the universal concept is completely and exclusively present in every object it refers to. It is therefore possible to unite all individual objects of this species into a single one (e.g., *the* oxygen) containing both the universal and the particular. In the organic world, on the other hand, a union of all individuals is meaningless. The individual offers a resistance that compels the universal concept to designate a series of attributes as inessential, to exclude them from rationalization.

In the Western tradition, the task of *classification* has been less important than that of discovering structural laws. This fact is related to the predominance of the mathematical natural sciences and the physical method of law. The general resistance currently offered to the exclusive dominance of the physical method—we have already mentioned this in the discussion of phenomenology—will undoubtedly benefit a gestalt perspective that reveals the meaning of species more profoundly than the schematic systems of biology were able to do.

Structural laws and structural forms are not fundamentally distinguishable, for laws vary with the forms, and every universal form is itself a law. The distinction is only that structural laws disregard the unity of form, and structural forms proceed from the unity of form. But laws and forms create a *gestalt hierarchy*, beginning with an idea of the highest gestalt of life as such and concluding with the most individual gestalt. Thus there is no definite boundary in biology between the universal and the particular. Structural forms and structural laws are ideal concepts; they vary so much in every individual that finally every living thing constitutes a special case—of which biological technology (e.g., medicine and cultivation) knows more than does pure biology.

The Heterogenous Methods of Biology. The significance of the individual in biology becomes even clearer when we observe the heterogenous invasions of the biological sphere by the methods appropriate to the sciences of law and sequence. Strangely enough, both methods begin with the same theory, the *theory of evolution*. This has led to the establishment of both laws of evolution and sequence contexts. In the first case, the theory of evolution is a grandiose attempt by rational, mechanical thought to subdue the biological sphere. It establishes a unity that encompasses everything from atomic life to the spiritual life of humanity. Yet the theory of evolution is a myth. It is an alternative to the Genesis myth, of course, but like that myth, it is inaccessible to science. The origin of life from the inorganic, the evolution of species through purely mechanical forces—these are conceptions that become impossible as soon as one grasps the gestalt concept. Life cannot be explained, for gestalt is a primary category that can be viewed only metalogically. A purely objective attitude cannot comprehend life. The creative element revealed by every living gestalt can be grasped only *re-creatively*, by empathetic understanding. Of course, the understanding of life is impeded by the fact that the dimension of life is prior to the dimension of consciousness. But life is, after all, that which comes to itself in consciousness; therefore, it is possible for consciousness immediately to grasp the element of life residing in itself. Without this act of grasping the gestalt of life, biology is dominated by an abstraction that keeps it from its proper object.

This gives us a new understanding of the view of nature held by Greek and German idealism: Aristotle's "entelechy," Schelling's "potencies," Hegel's "objective spirit," Schopenhauer's "objectification of the will"—all these express the *conceptual struggle* between thought and being. The understanding of the original existential element strives for expression in these concepts. They became fatal to science only when they were treated as causal principles of explanation, thus hindering real causal explanation. But that is not their meaning; it is a perversion of their meaning.

Just as biology has appropriated the law method—an appropriation that is only partly legitimate—the theory of evolution has generated the formation of sequence contexts. This theory stands within the individualizing series of the physical group of sciences; it is therefore directly related to paleontology and biological geography and is, in the proper sense of the term, *natural history*. Nevertheless, the sequences of species spoken of by the theory of evolution are not genuine sequence contexts, for species sequences (even including biological individuals, such as highly-cultivated sequences of stock) are not completely individual gestalts. They lack the mark of genuine sequence, the spiritual individual. Thus there is no logical contradiction involved in thinking of the various species simultaneously and in treating them in a merely geographical and individualizing way. But for genuine sequences, the time relation is constitutive.

ii. Psychology

The Autogenous Methods of Psychology. Psychology is the most controversial area within systematics and the theory of

methods. The controversy appears whenever the contrast between the psychic and physical dimensions is considered exclusive. For to assume this is to create a gap in the sphere of objects, a gap that is methodologically unjustified. One must
then choose either to give predominance to the material contrast between the natural sciences and the human sciences or to
give predominance to the methodological affiliation of
psychology with the generalizing sciences. Our own position
avoids this alternative. The inwardness that is the object in
psychology is already present in biology; in biology, understanding presupposes the self-comprehension of life. Indeed, we
have seen that the psychic and the biological dimensions must
be conceived as two sides of the one gestalt reality. From the
side of biology, this conception is primarily derived from the
fact of the individual biological gestalt; from the psychological
side, it is derived from the area of the unconscious life of the
soul, an area no dualism can grasp. When the *individual
power of formation* and the *unconscious life of the soul* coincide, the enigma is solved, and with it the one primary enigma
given in the very relation between thought and being. Thus we
are not speaking of an objective metaphysics of the unconscious; we are speaking of an understanding of internal
creativity. But this understanding fulfills the deepest intentions
of Western thought, which has always viewed internal productivity as a creative source.

From what we have said it follows, methodologically, that
the psychic dimension must be treated as the interior, conscious part of the biological dimension. The principle of all
work in psychology is internal intuition. Since biology is concerned with the preconscious dimension, introspection on the
part of life is not the source of individual cognitions. But in
psychology, all cognitions are made possible by introspection,
though all cognitions are not established in this way, for self-
observation, or introspection, is corrected and expanded
through an observation of others. But to understand the life of
other souls, one must first understand the life of one's own. A
methodological distinction from biology cannot be derived
from this, however. *Introspection* signifies a distinction between forms of reality, not between conceptual formulations.
That which is given, the psychic process, is subject to the com-

plete objectivity that characterizes biological objects. Subject and object are as separate in introspection as they are in biology; the re-creative act of understanding is directed only to the gestalt character of the soul, to the living unity of all processes, not to the individual objects of psychological investigation. From this it follows that the problems involved in formulating psychological concepts are analogous to those involved in formulating biological concepts.

The psychological gestalt laws correspond to biological gestalt laws. The prius of gestalt laws is the gestalt to which they apply; the establishment of these laws is thus always related to the living context of the entire gestalt. Psychology would have made more progress if it had originated as a descriptive gestalt science with the cognitive attitude of understanding instead of as an explanatory science of law. It would then have avoided the antithesis between psychophysical causality and psychophysical parallelism as alternative ways of explaining the relation between *body and soul*. It would have avoided the position of psychophysical causality by recognizing the intimate connection between the internal and the external in every living organism forming a gestalt unity, every organism for which a relationship of equivalent causality would be nonsensical. It would have avoided psychophysical parallelism by recognizing that the psychological gestalt is destroyed when its individual functions are coordinated with the biological ones. The total psychological gestalt and the total biological gestalt are coordinated as the internal and external aspects of the undivided, living gestalt, which is expressed in each aspect in a distinctive way.

Just as in biology, the individual side of the psychic gestalt leads to the formulation of generic concepts. These concepts can conform closely to biological concepts and can establish various species of the psychic dimension in the series of living beings, depending on the biological species. The origins of this procedure are perhaps found in the distinction between *animal and child psychology*. But so far, the procedure has been used very little, because subhuman psychic life is essentially inaccessible to us. More significantly, however, classification can expand into general psychology, but with one important difference from the formation of biological classes: here the individual is so significant that *type* replaces species. The species is a rule to which there may be exceptions, but these exceptions are regarded as degenerations. The type is an ideal that allows only individual approximations; seldom does an individual belong to only one type, as is the case with species. The concept "type" thus contains a much stronger resistance of being to thought than does "species." This explains the recent demands for the creation of a differential psychology and characterology. These demands are basically fulfilled in the theory of psychological types.

The Heterogenous Methods of Psychology. The distinction between autogenous and heterogenous methods makes it possible to do justice to the invasion of the psychic domain by the method of mechanical causality. The *associationist psychology* currently under attack and the use of *experimental methods* have the same right and the same limits within psychology as do the attempts to explain life in mechanistic terms. The psychological process also contains quantitative elements that can be isolated and subjected to special investigation. But the limits of this method are very narrow. Every psychic process belongs to an individual gestalt, a "soul," and receives its peculiar qualitative determinations from it. No element of the soul lacks this qualitative coloring. This corresponds to the situation in biology. But at the same time, it implies a powerful reinforcement of the individual tendency. To the degree that biological inwardness has psychic consciousness, individuality is strengthened. The *ego* is the complete separation of the individual gestalt from the whole of reality; it is the reflection of the individual upon itself.

Psychology has become the arena of a conflict over the concept *"soul."* On the one hand, there is the position that the soul is an objective, substantial reality to which one can attribute such qualities as freedom and immortality. In contrast to this objectification of the soul, the critical position reduces the soul to the synthetic unity of all psychic functions. Here as everywhere in the theory of the categories, dogmatic realism and critical formalism stand opposed. The metalogical concept of gestalt overcomes this antithesis. It shows that the concept "soul" initially meant more than a merely formal relation of unity; it meant the existential root as substance in the metalogical sense. On the other hand, the metalogical position rejects every objectification of the soul; it sees substance only in the living psychic gestalt. The explanation of the soul by associationist psychology is therefore impossible. The living productive unity of the psychic gestalt is the prius of all functions of the soul.

In psychology, the heterogenous sequence method stands alongside the heterogenous method of law. It is possible to arrange the structure and types of soul within a sequence context and to observe a development and transformation of the soul. Various changes have occurred, especially under the influence of the spiritual fulfillment of the life of the soul. Of course, one must rigorously distinguish a *history of the soul* from the history of the contents of the soul, which belongs to the genuine sequence science.

Psychology and the Human Sciences. The last remark leads to one of the most important philosophical problems of recent times, the problem of determining the relation between psychology and the human sciences. It is necessary to discuss this relationship in this second part, because both the independence of the human sciences and the possibility of the third part of our system depend upon our decision in this matter. By "spirit" we have understood the self-determination of thought as an existing thing—existing, that is, in the consciousness of living beings. From this it appears to follow that the sciences whose object is the life of the soul must be called in the *sciences of spirit*, or the human sciences, and that all human sciences are basically psychology. But this interpretation confuses the existential form of thought with thought itself. Psychology is the science of the existential form of thought, not of existing thought. The one science begins where the other leaves off. The one provides the container, the other the fulfillment. Spirit differentiates itself from all mere existents, including psychic existents, by subjecting itself to the form of thought, by striving to be valid. Spirit stands over and against all being with an unconditioned demand. But this contrast is much more radical than all the distinctions within the empirical sciences; it is absolutely fundamental. To obscure it is to deny that validity differs from being—the basic error of *psychologism*. When psychologism treats spiritual things as "higher faculties of the soul," then it can either establish that such phenomena are found in the intricate structure of the soul's life and that their origination and cessation, their waxing and waning, their combinations and contrasts must be arranged within the total context of the psychic processes; or it can seek the essence of spirit within these relations themselves. The first option is correct and necessary, but it does not discover the essence of spirit; the second is impossible and destroys spirit, whose peculiarity consists precisely in the negation of pure being and its contexts. Whenever this *"No!"* *of the spirit* against all immediateness of the purely given is ignored, the essence of the human sciences has not yet been grasped.

This clarifies the role psychology alone can play in relation to spiritual contents. The entire *psychology of culture* (including the psychology of law, of morality, of science, of art, and of religion) is a description of the forms in which these spiritual functions are psychically realized. But this says nothing about their content. Nothing is more disastrous than the confusion of psychic forms and spiritual contents, of the psychology of culture and cultural systematics. All psychology of spirit is heterogenous in relation to culture; it does not find the essence. But it is highly significant when it distinguishes the constant forms of psychic realization from the individual historical creations.

Wundt has provided psychology with a new task: *folk psychology*. This discipline is supposed to supplement and complete individual psychology. It presupposes that the psychic functions undergo transformations through the social community, transformations that cannot be explained by individual psychology alone. This fact is undeniable. But can it establish a folk psychology as an independent science? Even Wundt's work shows that it cannot. Of course, one can distinguish two elements in Wundt's psychology of peoples: history and the sociological theory of forms. But even the history of the primitives is not the psychology of peoples. The spiritual state of the primitives is no less unique than that of the most highly developed cultures. This state is one of *original spiritual creation* — somewhat different from the spiritual state of high cultures, but not unspiritual, not merely psychological. The treatment of primitive cultures from the perspective of folk psychology vastly underestimates the spirituality of these cultures, because of its individualistic, rationalistic standpoint. Naturally it is possible to ask to which psychological types peoples, ages, races, and even individual personalities belong and to ask on the basis of which natural talent they have attained their spiritual peculiarity. But a folk psychology in *this* sense does not belong within psychology; it belongs to the biographical division of history. On the other hand, folk psychology as a gestalt science can be realized only in sociology.

iii. Sociology

The Nature of the Social Organism. In the biological and psychological dimensions, the individual has attained its highest state of independence by achieving self-consciousness. But this development is misleading. The apparently isolated individual stands on the same existential ground, and is shaped by the same form of thought, as every other individual. This fundamental unity is maintained as isolation progresses. But it is maintained through the growth of a comprehensive gestalt, the *social organism*. Sociology is the science that is concerned with the inclusion of the individual within the comprehensive unity of the social gestalt.

A social organism appears only when the individual has developed into particularity. This has not yet happened in the dynamic sphere. The things that are completely subject to reason coexist in a disconnected way; all of them are *impenetrable*. Mutual inherence can emerge from coexistence only when thought is realized in things—in other words, when being opposes thought. Not that which is uniformly subject to a law, but that which individually shapes and distinguishes itself, is receptive to socialization.

Unlike biological and psychological gestalts, the social organism is not externally visible. It is perceptible only in its effects on these other gestalts. Therefore we must clarify its nature from both the biological and psychological sides, but so that we reveal its character as a gestalt in its own right, distinguishable from the visible gestalts.

Above all, we must avoid the impression that the social organism can be *explained* from biological or psychological motives. This common view is completely false. When there is a social unity, certainly there are individual motives, subjective interests, and pure relations between objects, all of which are important for social alliances. But one cannot explain this unity by these relationships, which already presuppose the existence of social unity. This is the one heterogenous invasion of the explanatory method of law; it is comparable to the attempt of associationist psychology and the theory of evolution to explain the soul and life by mechanical causes. The attempt to construct the social organism from the individuals in which it is realized always ends with the destruction of the organism. This is true in both the practical, political sense and the theoretical, scientific sense. The social organism is a gestalt — a gestalt that is perceptible only from the perspectives of biology and psychology, but that is nevertheless completely independent of these sciences.

The same creative act of life that, in biology, creates the independent individual organism also places this organism within an organic social unity. Here there are not two acts, one that posits the individual organism and another that posits the sociological relations; rather, the individuals are posited as related to each other, whether this relationship be one of community or one of conflict. Just as it is the same act that posits the interior and the exterior, the psychological and the biological gestalts, so it is the same act that posits these gestalts both as individuals and as sociologically connected. Some thinkers have used the notion of a common *organic substance*, which transmits itself through heredity and lives in all individuals within a group, to interpret this creative act. This notion is acceptable as long as one does not interpret it objectively, allowing it to become mechanistic through concepts such as "heredity." Here we must eliminate every relation of equivalent causality. The community is a *primary gestalt*, alongside the psychological and biological gestalts; community can occur with and in the causal processes, but it is always the presupposition of such processes and can never be explained by them. Community is a substance in the metalogical sense; it is a peculiar reality that is rooted in being.

The same considerations arise from the side of the psychic gestalt. However true it may be that the selfhood of individuals isolates them, the ego is still an abstraction when it is isolated.

The actual ego is always sustained by a comprehensive unity; it is simply impossible to separate an individual ego. Language is an especially important function of this comprehensive unity. The impossibility of isolating individuals is shown by the fact that even wordless thought and mystical absorption must use this social instrument of language in order to achieve consciousness. The understanding of other psyches also presupposes an originally given functional unity; this understanding cannot be explained by conclusions drawn from one's own psychic life. Moreover, all spiritual life is possible only to the extent that it is social. The social organism is thus the prius of all psychological social forms; it cannot be derived from them. Again, some thinkers have assumed the existence of a real *universal ego* or a mystical spiritual substance in order to understand this fact—with the same justification, and lack of it, as in the case of organic substance. If they conceive the universal ego or the spiritual substance (in the metalogical sense) as an independent, given gestalt, there can be no objection to this notion; but as soon as they posit an occult object instead of a creative gestalt, the notion must clearly be rejected.

Thus the metalogical method understands the nature of the social organism against three positions: (1) against the attempts of a rational, mechanical *explanation*; (2) against logistic *formalism*, which speaks merely of a logical unity; and (3) against a rational *objectification* in the material, substantial sense. The metalogical method allows us to view the existential act of creation, the act that realizes itself in the logical form of unity. Only this method can do justice to the nature of the living gestalt.

The Objects and Methods of Sociology. The objects of sociology are the *structural forms* and *structural laws* of the social organism. Sociology has been defined as the science of the forms of socialization. This definition points to the fact that the social gestalt is visible only in its psychological and biological expressions. But it creates the impression that the social organism arose through the activity of its members. It is therefore better to say that the objects of sociology are the structural forms and structural laws of the social organism, not the forms of socialization.

The foundation of all sociology is the theory of the essence and categories of society, or *social philosophy*, which we have treated in the previous section. The universal social laws that are valid for every social community are based upon this discipline. These laws are applicable to various kinds of social formation, from the most transient contact between two persons to the unity of the human race, and beyond that to the ideal cosmic social structure. To the extent that they enter a social form, all biological and psychological functions can become the object of the *doctrine of sociological forms*. This is also true for the spiritual functions. Nevertheless, just as the psychology of religion, of art, and of morality do not make pronouncements about the essence and significance of the objects within these functions, but merely establish their existential forms in the psychic life, so the science of sociology does not replace the human sciences. According to one view, sociology seeks to assimilate spiritual things (which are always biological, psychological, and sociological things *as well*), just as psychology and, before that, biology sought to do. *The human sciences are equivalent to sociology*: this formula has been influential from Comte to the present and has made the classification of sociology so difficult. Sociology can be understood only when this error has been overcome. The task of sociology is to extract the constant structural laws from the contents of the social life. Thus in every sociological investigation, one must ask whether the process under consideration is an individual spiritual creation or a sociological form for every possible spiritual creation. This procedure is not without difficulty, because it is very easy to infer the formal character of a content from its frequent appearance. It is therefore important that sociology perform this task by disregarding the objects of the human sciences and that it distinguish the constant forms of the social life from its variable contents.

The *method* of sociology is completely determined by the gestalt method in general. Here we find the autogenous gestalt methods: the establishment of structural laws and structural forms. We find both the heterogenous invasion of explanation in terms of mechanical causes and the arrangement of sociological types and stages of development within sequence contexts. The latter forms a transition to the methodology of history. But the two must be distinguished: the sociological theory of types is concerned with the development of the naturally given, structural forms of society, and historical investigation is concerned with the creative contents with which these forms are filled. An inspection of the social functions shows, of course, how difficult it is to separate the two.

The Social Functions. There is a series of functions that not only can assume social form, but that are essentially social — functions through which the social organism lives and expands. This series includes, first of all, those functions in which the association of individuals within the social organism appears: the *spiritual association* in language and writing, the *spatial* association in commerce. Second, there are those functions in which the social organism *maintains* itself and *expands biologically*: social hygiene and social economy. Third, the functions of the *political development* of the social organism: internally, the administration of justice and of the state; externally, defense and offense. And, fourth, the function of *spiritual development*: social pedagogy.

These social functions would undoubtedly be the most important objects of sociology if it were possible to regard them as empirically given functions. But that is impossible. None of these functions is merely given; each is also proposd as a task, as the object of a conscious formation. *Conscious formation* can, however, be either technical or spiritually creative, or both. It follows that the content of all the social functions belongs to areas other than sociology. Thus language belongs to philology as far as its content is concerned, but to sociological technology as far as its social application is concerned; most of the others belong to the practical human sciences as far as their goal is concerned, but to sociological technology as far as their execution is concerned. The only task remaining for sociology is the theory of sociological forms, which is valid for these and all other functions.

Nevertheless, the social functions have a sociological presupposition. This explains the attempts to make *empirical* investigations the foundation of sociological technology. According to these attempts, just as the foundation of medical work is laid by biology, so the norms of sociological technology are based upon the knowledge of the laws of social life; and just as the goal of medicine is the health of the body, so the goal of social technology is the health of the social organism. But it is impossible to maintain this view, for the empirical sciences cannot establish goals for social functions. The establishment of goals is a matter of spiritual creativity; it is dependent on the fundamental ethical attitude. The concept "health" is completely inadequate for this purpose, for under some circumstances, spirit is opposed to the will toward biological self-preservation. But the positings of all other goals, and of their appropriate technical means, are directed to the determination of the fundamental goal. These means are not arbitrary, of course; as in every technology, they depend on the material, and the material of social technology is sociological reality. But what should be made from this material depends on the positing of the ultimate goal; this is the task of the human sciences.

It is also impossible to separate the sociological material from its connection with the technical forms realized in history, in order to achieve knowledge of the universal laws of social structure. There are no such pure structural laws standing in isolation from the sociological theory of forms. The sociological organism is always a *self-positing gestalt*. It posits itself through spirit and technology. This self-formation is limited only by the sociological forms; but sociological forms are not social functions. Therefore the social functions belong, on their normative side, within the practical series of the human sciences, and on their technical side, within the sciences of sociological technology.

c. The Universal Significance of the Gestalt Method

The invasions of the method of the law sciences have been evident in all areas of the organic group. We have established both the justification for and the limits of the incursions of the law sciences into the gestalt sciences. But there is also an opposite tendency. The gestalt method claims *universal significance*, attempting to absorb all other areas. This tendency is continually nourished by the depths of the primitive consciousness, with its objectively mystical intuition of the whole of reality. It finds its scientific form in the aesthetically illuminated cosmologies of idealistic philosophy. The idea sometimes assumes biological form (*world organism*, in Bruno and Schelling), sometimes psychological form (*panpsychism*, in Leibniz), and sometimes sociological form (*the realm of spirit*, in Neoplatonism); yet these forms are usually mixed (*world soul*, in Bruno and Schelling; the *series of spirits*, in Leibniz). In particular, biological and psychological conceptions are readily applied to the celestial bodies (Plato, Fechner). From the perspective of the physical sciences of law, all these representations are fantastic, being on the same level as the myths of primitive religion. Though correct in itself, this criticism of every fantasy of nature tends to overlook three things. First of all, on the basis of the theory of the categories, the organic is indeed the prius of the inorganic, not conversely. The idea of a universal gestalt within which the physical processes occur is a necessary idea. In the second place, the mechanistic world view must itself presuppose the uniformity of the whole within which the mechanical processes occur. It is no accident that the mechanistic cosmology originated in Neoplatonic nature mysticism and initially sought to prove the harmony of the cosmos. The idea of the *cosmos*—that is, the idea of the uniformity and gestalt character of the world—is tacitly presupposed by the unhistorical, mechanical explanation of nature. And third, one must bear in mind that epistemological idealism maintains that the unity of the conscious gestalt with the existential gestalt is a necessary precondition of knowledge. If this presupposition is correct, then the spiritual structure (in the sense of "consciousness in general") becomes the interior aspect of the structure of reality (in the sense of the "the universe"). Even from this point of view, the idea of the cosmic gestalt becomes the demand of thought.

Of course, an objectively mystical conception of the world cannot be established on this basis. We refer to the fact of the ultimate *positiveness of being*, of the fundamental givenness of a cosmic gestalt—a fact that must be thought as the ultimate prius of all mechanical, organic, and psychic processes, the fact of the original creation, which can be grasped only when thought stands reverently before the abyss of pure being, of the absolute position. Like the comprehension of life, the self, and society, this act obviously implies a metalogical understanding. But unlike the other gestalts, the cosmic gestalt is not an object of experience. It is only presupposed; it is not given. For although the creative principle reveals itself within the world forms in the individual gestalt, it can still realize itself in the totality only through the infinite sequence of new creations. Thus it demands the inner infinity of being. The gestalt nature of the cosmos is eliminated by the sequential nature of the infinite act of creation. In this way, the dialectic of thought and being is revealed in its ultimate depths: the gestalt is transcended by history.

2. The Technical Sciences

a. Foundation

i. The Organic and the Technical Sciences

The gestalt sciences are divided into the organic and the technical sciences. In the organic sphere, the goal is inherent in being; in the technical sphere, it is posited from outside, originating in the subject. Usually the *contrast between organic and technical* is regarded as exclusive, and the two concepts are made the symbols of opposing attitudes toward life. One could say that the organic originates in being, the technical in thought. No doubt this antithesis is present in the history of the Western spirit. It is based upon the rational, world-shaping will in the service of which Western technology stands; it has received its acuteness from the powerful heterogenous attack to which the subjective positing of goals, in alliance with the method of physical law, has subjected all areas of the empirical and the human sciences. But the antithesis does not exist in the autogenous sphere of the technical and for the method of technology as a science.

Technology is the shaping of reality according to a goal. The technical sciences view their objects as *goal gestalts*. Every part of a technical gestalt is a member; it loses its meaning outside the whole to which it belongs, reverting to the sphere of mere physical being out of which technology had raised it. Accordingly, in technology we find the cognitive goals of the organic sciences: structural laws and generic concepts, types and heterogenous sequences. The difference is only that in the organic sphere, the goal is present in being, and in the technical sphere, the goal is impressed upon being from outside. But even this contrast is partial.

Technology is related to the different areas of being in very different ways, of course. Because the goals of the organic group are inherent in being, technical activity must destroy the immanent goal of an organic gestalt when this activity posits other goals. For example, technical activity can make lumber from trees only when it destroys the tree as a gestalt and makes the material "lumber" from it. In the organic group, technology can avoid destroying gestalts by realizing only those goals that correspond to the inner tendencies of these gestalts. If we call all of these tendencies "development," including protection and preservation as its negative presuppositions, then we can call technology in the organic group "the technology of *development*."

Physical things are not inherently related to goals. For them, a goal is something alien, something that has nothing to do with their inner tendencies—something, indeed, that contradicts these tendencies. Accordingly, physical technology can be called "the technology of *transformation*." Obviously, this form of technology always depends on the native laws of its forces and materials, but it compels these forces and materials in a direction that is foreign to their nature; it is therefore in constant conflict with the natural direction of its objects, in accordance with the saying that the elements hate the construction of the human hand. Whenever technology is victorious, it shapes the image of reality.

Modern natural science has not produced technology (especially physical technology) as an accidental by-product. From the beginning, natural science has evolved in close alliance with the technical will. Since the late Middle Ages, natural science has been guided by the ideal of the *technical control of the world*, and its greatest triumphs were always technical victories. The tacit superiority the so-called pure sciences nevertheless enjoy is based less on the fact that they are fundamental than on the fact that the Greek ideal of theory has, through humanism, never allowed the actual rank of the technical sciences to be recognized. But the real attitude of the modern Western spirit is undoubtedly that the technical sciences are the proper end of scientific endeavor.

Because of the very nature of the historical sciences, they have no relation to technology. Indeed, the peculiarity of the creation of the unique is that it is essentially *free from goals*. Being is "older" than goals. And where being has divested itself of all rational bonds, goals have no place. Thus we find the conflict between thought and being in the three main divisions of the technical sciences: in the physical group, the goal as an alien law forces the existent into its own channels; in the organic group, being assimilates the goal, allowing itself to be guided by this goal only insofar as the goal is consistent with the inherent goal of being; and in the historical group, being stands in goal-free, creative uniqueness, exempt from every technical influence.

ii. The Method of the Technical Sciences

The methodology of the technical sciences corresponds to that of the organic group. There are, however, notable methodological differences between the sciences of the technology of transformation and those of the technology of development. The methods of the former are nearer to the physical methods, those of the latter, to the historical methods.

Unlike the organic gestalts, the products of the technology of transformation have no basis in being. These technological gestalts therefore resemble the structures of the thought sciences. They are formed from outside—not through definition, of course, but through the actual transformation of existential reality. Thus the deductive and rational elements of this group's methodology are especially prominent. Yet we are concerned here with gestalt concepts. For the prius of every establishment of a law in technology is the goal gestalt that is produced by description. Even the categorial view does not disagree with this. Indeed, productive causality is missing from physical-technical gestalts with immanent goals. Technical gestalts are not substances in the primary sense, yet they are not without substance; they are substances in the secondary, dependent sense. They originate in the primary organic substances by which they are posited, substances that provide them with part of their own substantiality. Physical technology is the *rational formation of organs* by spiritually conscious creatures. Through this form of technology, the technical gestalts have a derivative, secondary productivity. These gestalts form a closed causal system, which is free from the immediate physical process and which gives a definite, preformed direction to every physical cause influencing the gestalt. But this fact is methodologically important; it necessitates the placement of physical technology within the gestalt sciences, in spite of the fact that from the categorial point of view, the *gestalt character* of physical technology is only secondary.

The material and product of the technology of development is the organic gestalt. Thus no one can doubt the gestalt character of these sciences. But just as the technology of transformation is closely related to the physical sciences, so that of development is related to the historical sciences. The praxis of the technology of development is always concerned with individualized *particular gestalts*. Medicine and cultivation, pedagogy and economics, administration and diplomacy — all these sciences are directed to concrete gestalt individuals. A science of these things must therefore be so constituted that it can do justice to the particularities of the individual gestalt. But the individual is grasped by empathy and understanding. Organic technology develops its gestalts on the basis of re-creative understanding. It is therefore largely constructive in its procedure; it has the convictional nature characteristic of historical understanding. Yet it is a gestalt science, because it grasps the individual, not for its own sake, but for the sake of the goal it posits for all individuals: the goal of the ideal, developed gestalt, which is itself determined by the norms of the human sciences.

The methodological distinction between transforming and developmental technology is especially important for sociological technology. Here one can observe a continuous heterogenous influence by the basic attitude of the physical sciences. The self-formation of the social organism is conceived more in the sense of *rational* construction than of understanding construction. The given gestalt is not developed; it is destroyed and relegated to the physical sphere, to be reconstructed in the manner of transforming technology — an action that is impossible in itself and that has a heterogenous right only within certain abstract considerations. On the other hand, organic technology must sharply distinguish the meaning of developmental activity from the meaning of transforming activity, not only in sociology, but also in the other organic spheres. Today, in reaction to the mechanical perversion of the technology of development, the concept "the organic" sometimes symbolizes a merely *vegetative* ideal, especially in the sociological sphere; but this view forgets that the development of the social organism proceeds only through conscious activity that is subject to spiritual norms. But these norms cannot themselves be derived from the organic sphere; they are creations of the spirit, and they shatter the closed gestalt in favor of the productive sequence series. They create history from the organic. Thus the technology of transformation rises from the physical into the organic sphere, and the technology of development rises from the organic into the historical sphere. But in all its parts, the science of technology is a gestalt science.

iii. Technology and Spirit

The technical sciences are not only gestalt sciences; they are also formative [*gestaltende*] sciences. They form being, whose nature and laws they seek to know. They are productive, like the human sciences. And like the latter, they both *posit being* and comprehend it. This could lead one to believe that the technical sciences and the human sciences belong together, that technology is a creative spiritual function. This view would find support particularly in the technology of development, where technical acts of formation are frequently codetermined by social or legal norms. But this view is erroneous. In technology, spirit does not determine itself; it determines being, which is alien to it. Goal gestalts are posited, and they are subject to the laws of the empirical sciences. In technology, the positing of goals is creative, in the sense of the human sciences. In this respect, the technology of development is no different from the technology of transformation; though its positing of goals may be determined by morality and law, its execution depends on the existential structure of its gestalts.

The following distinction might clarify our understanding of the proper sense of creativity: in the sciences of thought, the object of knowledge is *found*; in the pure empirical sciences, it is *discovered*; in the technical sciences, it is *invented*; and in the human sciences, it is *created*. These terms correspond exactly to both the cognitive attitude and the procedure of knowledge within the various areas.

The differentiation of the technical sciences from the human sciences is also important for the history of technology. Should technical gestalts be placed, like spiritual creations, within a genuine sequence context, and is the *history of technology* a necessary component of cultural history? This is a difficult question, for both technology and culture contain two elements: a spiritual, normative element and a technical, existential one, or creation and invention. When we consider technically formed reality, we are dealing with a continuation of the heterogenous sequence series of the physical sciences. We encountered this sequence series in all the organic sciences; it is present throughout the technical sphere; and it is increasingly connected with the autogenous investigation of history in the technology of development. On the other hand, if we consider the powers that posit goals (i.e., if we consider the spiritual functions), then the history of technology becomes part of universal, autogenous historiography.

One clear distinction between invention and creation is that invention is in principle *subject to obsolescence*, while creation is inherently infinite and can become obsolete only on its technical side, never on its creative side. Therefore, the history of the most ancient cultures has novel significance for every present, but the history of technical transformations (even those of recent times) has no significance for the present, when it is considered only from the technical standpoint.

iv. Technology, Craftsmanship, and Applied Art

"Every object has its science." This saying points to a methodological problem, that of where "art," or craftsmanship, ends and science begins. In principle, every object can undoubtedly have its science—that is, every goal, however insignificant, can lead to a systematic investigation of the means. But in reality, scientific reflection is largely replaced by instinctive, acquired, and inherited praxis. And praxis retains its position even when science has been in effect for the longest time. Mechanical engineering is scientific theory. But actual mechanical engineering is impossible without praxis. Even in mechanical technology, concrete reality, or the individual gestalt, always contains a profusion of irrationalities that are accessible only to instinct and practice and that remain alien to pure thought. On the other hand, science strives to free praxis from contingencies and inadequacies and to rationalize it as completely as possible. Thus *science and craftsmanship* both cooperate and contend with each other in all technical activity. But the fundamental difference can be stated in this way: science is present whenever there is a coherent and exhaustive theoretical foundation based on empirical knowledge; craftsmanship is present whenever a sum of accidental experiences and observations is transmitted to customary activity. Thus the distinction is based on method, not on objects. Every technical activity contains an element of craftsmanship, and every such activity can be based on scientific grounds.

Alongside the element of craftsmanship in technology stands the artistic element. The problem is already given in the term "art": taken in the sense of a faculty, it first of all expresses only technical ability; but it has also signified a special function of the spirit, the aesthetic function. Art, in the sense of a faculty, belongs to the practice of every cultural activity. There is an art of research, of preaching, of painting—all of which belong within the sphere of the technology of development. Insofar as it shapes being, technology can combine with art, leading to the peculiar dual phenomenon of *applied art*. Applied art appears only in the technology of transformation. It is true that the attempt has often been made to regard even the products of the technology of development—a highly cultivated biological gestalt, a complete social *organism* (for example, the state)—as "works of art"; but this use of language is figurative. Organic gestalts are determined by their own goals. They can have aesthetic significance only insofar as they are perfect natural phenomena; but they can never be works of art, for the artistic character is foreign to the immanent goal of the material. Consequently, there is applied art only in the technology of transformation, the technology that is essentially concerned with the shaping of alien material.

The most important area of the technology of transformation is *architecture*, which combines science and art. Both technology and aesthetics are interested in architecture. This creates numerous problems for both praxis and aesthetic theory. From the point of view of scientific systematics, this presents no difficulties, because the technical and the artistic views are completely independent. The technology and aesthetics of architecture are separated in the same way as are the investigation and aesthetics of nature; they are closely related only in praxis.

v. Value and Goal

In the technical sciences, the object of knowledge is the goal gestalt. But a goal is posited subjectively, even when it agrees with the intrinsic goal of the object, as in organic technology. But what kind of goal constitutes the technical gestalt?

The goal is posited by the cognitive subject; it resides in this subject. The goal of technology is man. But this statement is insufficient. Man is actualized biologically, psychologically, and sociologically; technology serves all three of these aspects. But man as a biological and psychological individual is an abstraction. Only man within the social organism is real. All technical goals ultimately serve the formation of the social organism, and the technical sciences deal with reality insofar as it can be formed for the goals of the social organism. Reality is limited from a *standpoint*; it comes into consideration only insofar as it can be transformed in the direction of this standpoint. Thus, in relation to reality the technical goal is a subjective standpoint, and the subject is the social organism and its members.

Here we encounter the pragmatic point of view for the second time. The first time, it confronted us as the principle of selection for the formation of heterogenous sequence contexts. What appeared there as the principle of selection now appears as the goal-positing principle: man as a living being, or concretely speaking, man in the social organism. The duality of objective and *pragmatic* elements within the formation of heterogenous sequence contexts is derived from this. Alongside the universal goal of knowledge, there is one that is limited by selection and purpose. But is it possible to have more than a subjective justification of the pragmatic element? Is the pragmatic goal merely subjective? Could any other goals be placed on the same level with it, or is there even a scientific foundation for it? Is the pragmatic goal a value? It is obviously a value for the subject who has the goal—the value life has for everything living. But there are various living beings, and each one claims this value for itself. What we seek is not a subjective value, but the objective, unconditioned value: absolute value. Only this value can justify the pragmatic positing of a goal.

The human sciences determine value. Only these sciences can give a scientific justification for the pragmatic positing of a goal. But one thing is crucial for answering this question: the fact that the technology of development, especially in the sociological sphere, is itself already subject to the positing of a normative goal, that technical activity here no longer serves merely a pragmatic goal, but serves the unconditioned goal. The social functions are subject to the norms of the practical series within the human sciences. Thus in the technology of development, the goal itself becomes value, the pragmatic becomes the valid.

b. The System of the Technical Sciences

i. The Sciences of the Technology of Transformation

We distinguish between two main groups of the technical sciences: the sciences of the technology of development and the sciences of the technology of transformation. Within these two groups, it is useful to follow the example of the individual empirical sciences, though it is impossible to do so in a strictly systematic way. There can be no self-contained system of the technical sciences, because new goals continually appear and new means are constantly discovered. Nevertheless, the example of the pure empirical sciences provides a firm basis for organizing the technical sciences; a classification according to goals would flounder.

The technology of transformation can be further divided in accordance with the two major groups of the physical sphere: in the generalizing series, physics is related to the technology of forces, materials, and tools; in the individualizing series, it is related to the technology of the production and transportation of material and to the technology of construction. In the first group, technical tools are created; in the second group, the surface of the earth is permanently transformed.

Thus *technical mechanics and dynamics, chemical and mineralogical technology,* are extensions of pure mechanics, dynamics, chemistry, and mineralogy. As a transitional area, *pharmaceutics*, which itself forms a transition to biological technology, is the extension of organic chemistry. *Mining* is the extension of geology; the *technology of commerce and of construction* are extensions of geography, being dependent on it but also determining it.

ii. The Sciences of the Technology of Development

Biological Technology. We return to the three major organic sciences, beginning with the consideration of biological technology. Life is the object of technical activity in two ways: it is protected and it is advanced. This distinction yields two basic forms of the science of biological technology: healing (including hygiene) and cultivation. Indeed, these two forms are distributed over the biological series (plants, animals, and man) in such a way that at first cultivation has priority and in the end healing does. The more that which is individually formed predominates, the less biological technology is able to form the inherent powers of life and the more it must be content to support these powers. Thus, the *cultivation of plants* stands next to *plant medicine*, *animal breeding* next to *veterinary science*, and *gymnastics* next to *medicine*. Naturally, the points of transition between these sciences are blurred: healing and cultivation, protection and advancement, are mutually determinative. But the division of labor is clear.

In considering medicine, one should distinguish medicine as a member of the system of the sciences from medicine as a field within an academic faculty. The medical faculty occupies a place between pure and technical biology. But biology is not medicine. Medicine in the proper sense of the term is only present when the goal of preserving life determines scientific work. Medicine is a science within the technology of development. In ordinary language, the technical implies the *mechanical*; there is doubtless a mechanical element in all technology. For economic reasons, the relation between means and goal requires an extensive mechanization of the technical process. But even in the physical sphere, mechanization is a secondary matter. What is important in both theoretical and practical technology is not the mechanical element, but *"ingenuity,"* or productive understanding. Naturally, this is especially true in the organic sphere, where the individual as such is significant and where false mechanization therefore leads to destruction. But technology includes both elements, for both of them investigate goal gestalts that are subjectively posited. Medicine is not mechanistically interpreted, but technology is freed from a mechanistic misinterpretation.

Protection and advancement are both based upon preservation. The technical activity that is directed to the pure preservation of life is *economics*. We have not distinguished among plant, animal, and human economics, because economics is always a function of either protection or advancement, and preservation for the sake of preservation is not a distinct goal. It is different when man is no longer considered merely as a biological organism, but as a free subject separated from the immediacy of nature by his reason, a subject with numerous needs whose satisfaction is a rational task. Economics is the science of the rational satisfaction of needs. Economics as a science appears only when there is an economic subject. But man as a rational being is always present as an economic subject, not just as a natural being.

Economics is either individual or social economics. The former is related to the biological sphere, the latter to the sociological sphere. *Individual economics* is based on the anthropological part of the biological sciences in particular. Individual economics includes all economic doctrines and doctrines of trade whose object is trade for the purpose of satisfying individual needs. Essentially these include *agriculture, craftsmanship, industry,* and *commerce*. The facts that the majority of these trades have been sustained by subsidiary ones and that all of them together constitute economics as a whole do not prevent us from considering them from the perspective of private economics. These theories are sciences of biological and anthropological technology.

The heterogenous sequence series of the physical sciences continues in the biological sphere as the geography of plants and animals, as anthropological geography, and as the history of evolution. The technological correlate of these areas is the formation of the soil from the biological perspective, or *agriculture* (including *forestry* and *gardening*, but excluding the corresponding doctrines of trade, which belong to economics). Even animal husbandry and breeding belong to this group; on the other hand, the technological point of view does not do justice to man here, for human migrations and settlements are determined by fate and freedom, not by an imposed technology. Among the areas we have mentioned, gardening is especially interesting. This is because it is partly an extension of architecture; like architecture, it contains both aesthetic and technical elements.

Psychological Technology. Like biology, psychology can be the object of technology in two ways: as an object of protection and as an object of advancement. Thus we must consider both physiological psychology and cultural psychology. Physiological psychology is related to both *psychiatry* and *psychotechnology*, which we will define as the theory of the development of the highest possible psychic power. Both are comparatively young, as sciences: through psychoanalysis, psychiatry has attained a significance far beyond the area of medicine; psychotechnology extends beyond the realm of science, partly under Indian influence. As praxis, however, both forms of psychological technology are ancient. They have found their highest perfection in the religious practices of confession and ascesis. Insofar as they are scientifically investigated in practical theology, we can classify these practices as theonomous forms of psychological technology.

The technical science corresponding to cultural psychology is *pedagogy*. It contains both the protective and advancing elements. Because pedagogy corresponds to cultural psychology, its goals depend on the norms of cultural activity in general—that is, on the human sciences. The theory of pedagogical goals falls within ethics and the theory of community; pedagogy itself, as the science of the spiritual guidance of souls, belongs within technology. This is not to disparage pedagogy, however. Technology is $\tau\varepsilon\chi\nu\eta$, "art"—not in the aesthetic sense, but in the productive sense, in the sense that it forms reality. The fact that the pedagogical relationship is always a social relationship is no more reason for the classification of pedagogy within the sociological sphere than it was in the case of economics. Social pedagogy corresponds to sociology. This classification is determined, not by the fact of socialization, but by the goal. In individual pedagogy, this goal is the guidance of the individual soul. Besides, pedagogy does not necessarily socialize. There is also self-pedagogy, the art of developing oneself and one's talents. This discipline contains all methods of training in the spiritual and technical functions, as well as the technical elements of the functions of the spiritual individual—that is, the technical elements of art and science.

Sociological Technology. We come now to the third group in the organic sciences, sociology and its corresponding technical disciplines. We encountered the sciences of sociological technology when we discussed the social functions in our previous treatment of sociology. For the functions of the social organism are all realized through technical activity. Thus we have the sciences of rhetoric and political journalism, commerce, administration, diplomacy and warfare, social economics and social hygiene, and social pedagogy.

From antiquity until the late Middle Ages, *rhetoric* was a well-cultivated science. It was the Greek Sophists who first perceived the power of the word over objects. With modern empiricism, the object achieved victory over the word: for empiricism, the power of speech lies in objectivity rather than in artistic form. From the time of empiricism until the present, only religious speech has received systematic attention. *Homiletics*, as rhetoric on a theonomous foundation, is a part of practical theology. It is still possible that, with the conversion of the spirit and its rejection of rational empiricism we have mentioned so often, rhetoric will regain its importance. So rhetoric belongs to the sciences of the technology of applied art; it combines technical and aesthetic elements.

Political journalism is concerned with both overcoming the spatial divisions of the social organism through the written word and forming a collective will despite these divisions. *Commerce* shows how persons and things can overcome space in the interest of the general social life. Both of these sciences must take elements of the technology of transformation into consideration.

Administration, *diplomacy*, and *military science* are obviously sciences of sociological technology. But *social economics* does not appear to belong to this group of sciences;

even the name itself has not been established. "National economics," which is still the designation most often used, is reminiscent of the mercantilistic theories; but it is too narrow to cover the entire field, for nations and states are not the only bearers of socioeconomic processes. The term "political economics" refers to the normative element in the positing of economic goals, but it goes beyond the strictly economic area into the political domain. The series constituted by private economics, national economics, and world economics is not a homogeneous series, because private economics is a methodological abstraction from all forms of communal economics. This is why we choose to distinguish between individual economics and social economics. But even this terminology can be misinterpreted. There is the danger that it could be conceived in the sociological sense. But social economics is the doctrine of a social function, and the sociology of economics is a part of the doctrine of sociological forms.

There are two material problems involved in the terminological one: the problem of distinguishing economics as a technical science from the pure empirical sciences, and that of distinguishing it from the human sciences. The first distinction is necessary because the economic process is conceived, especially by liberal economic theory, as a purely objective reality whose laws are established in the same way that concepts are formulated in the natural sciences. Now, every technical product follows the laws of the material from which it is formed. Thus social economics necessarily follows the laws of universal sociological form. But these laws are not economic laws. *Economic laws* develop only when the effective technical will, which is partly conscious and partly unconscious, has created economic institutions that are subject to these laws. Economic laws are laws of the economic social functions that are technically formed; they change when goals and means change. This is analogous to the situation in the other social functions, such as warfare and diplomacy. The idea of a "pure" economics is an impossible abstraction. Economic activity is the technology of social development. Even the liberal theory presupposes that both a definite goal of the social life and definite technical means are given; in this, it is no different from any other theory. It is different only in the goals it posits and in its statement of the correct means for realizing these goals. Once both goals and means are recognized, laws of the liberally administered economic function—but not empirical laws of economics—follow as a consequence.

The distinction of economics as a technical science from the human sciences is necessary when the political character of economics is emphasized. Economic rationalism and *utopianism* try to define the economic process exclusively from the perspective of the ethical, legal idea; thus they overlook the inherent lawfulness and limits of every means, however perfect. But even this constraint differentiates economics from the human sciences, making it a technical science of being. The fact that social economics belongs within the social functions is decisive for distinguishing it from both the empirical and the human sciences. For by this fact, the positing of economic goals depends on the positing of common goals, or the political idea; and the economic means depends on the social forms that are effective in all social functions.

Social hygiene deals with the protection and development of the social organism on its biological side. This science is concerned with such important problems as population growth and eugenics. Even here, there are no purely empirical laws; everywhere in this domain, the positing of goals, on the one hand, and sociological form, on the other, are decisive.

Most of the sciences of social technology that we have mentioned are also called *"political sciences."* The reason for this is that the legal community, the state, is easily the most important community involved in positing the goals of social technology. This term is admissible if it does not lead to the confusion of political science with the philosophy of the state and the normative doctrine of the state (or politics), both of which belong within the human sciences. But the term is not entirely appropriate, because the state, although it is effective, is not necessarily the subject of most social functions, and some functions are still inclined to free themselves from the state.

Statistics has a practical and intimate relation to the sciences of social technology. Yet it is not a science; it is a method that can and must serve all the sciences, though it has especially important functions in the area of economics. For in economics, more than in any other area, the description and supply of material depend on numbers. But all the other empirical sciences, pure or technical, also use data that are subject to statistical count. Statistics does not have its own area within the empirical sciences; it belongs to the methodology of science.

Finally, *social pedagogy* contains both technical and ethical elements. The doctrine of community shows which cultural goals the social organism ought to realize; social pedagogy finds means to realize these goals. On their technical side, the problems of popular education and training belong to social pedagogy, which, as the last of the sciences of social technology, forms the transition to the normative sciences themselves.

C. Third Group: The Sequence Sciences

1. Foundation

a. The Spiritual Individual

The positing of *individuals* expresses the resistance of being to thought. The more individualized a class of reality is, the more filled with being and real it is. Thus knowledge grasps being most completely when being is least subject to the forms of thought, when it most strongly resists them. The being with the most reality is not the most rational being but the most irrational being. The most profound grasp of being is the self-conscious renunciation of the attempt to grasp pure being; this is not a simple, naive renunciation, but a conscious renunciation born from the supreme struggle for form, a renunciation that breaks through the form—indeed, the form of the renunciation is precisely this breakthrough. Thus the gestalt breaks out of the rationality of law and is revealed as the supporting substance of the reality that is subject to law; thus the creative sequence breaks through the self-contained gestalt and is revealed as the living content that strives for realization in the gestalts. Only in the creative sequence is being grasped as living being; only in the creative sequence does knowledge achieve full concreteness. Even the gestalt is still an abstraction: it abstracts from the "self" that supports the gestalt, from the absolutely particular that gives substantial character to every individual gestalt. The science of sequence is therefore the true *concrete science*. It presupposes the general gestalt and works with general gestalt concepts, but its goal is the individual gestalt, not the general one.

In the physical group, there are no completely individual gestalts, because there are no genuine gestalts. The heterogenous gestalts are not essentially individual; they are individualized by the construction of pragmatic contexts. Their individuality is accidental, not substantial. In the gestalt group, everything is individual, for there are only individual gestalts; but these individualities are too indeterminate in their confrontation with thought to resist subjection to the species or the type. Even the ego, which is the reality farthest removed from the universal, is a universal gestalt; it is grasped by laws of psychological structure. Yet the ego introduces a new principle into the gestalt group. According to the dialectic between thought and being, complete separation from the universal is possible only when the universal enters the particular, when the particular itself becomes a universal. But the universal that has come into existence, or existing thought, is spirit. The *complete individual* is therefore the *spirit-bearing individual*. Only a science that is concerned in an autogenous way with individuals can be the science of spiritual individualities. This science is history.

In the physical group, individuals are differentiated by the formation of pragmatic frameworks that to some extent have an objective, though insufficient, basis in pseudogestalts and gestalt compounds. This perspective is continued in the organic-technical group. But here we find a new element, the sequence series resulting from the development of the gestalts: natural history and the history of technology. The differentiation of individuals in these areas is determined by the individual species characterizing the gestalts. These sciences present sequences of species and types but neglect pure individualities. In the historical group, it is just these individualities that are the object of consideration; this group deals with particular psychic and social gestalts in which the universal lives as spirit. Here individuals are differentiated in two ways: on the one hand, the particular spirit-bearing gestalt is an individual gestalt that is grasped by concepts that describe the gestalt; on the other hand, the description of the particular gestalts occurs for the sake of the spirit they bear. Thus every historical presentation combines *description of gestalts with comprehension of spirit*. The distinction of history from both the heterogenous sequence series and the history of spirit rests on this duality. The description of gestalts gives history its existential basis, its connection with the other empirical sciences, whose elements it completely assimilates. The comprehension of spirit makes history the foundation of the history of spirit and thus a presupposition of the human sciences.

b. The Principle of Selection in History

Compared with the other sciences, history is concrete. But it is not absolutely concrete. Everything existential, everything individual, is intrinsically infinite. But thought exists in finite forms; knowledge establishes finite contexts. Even history is abstract, in the final analysis. It selects elements that are important for the context. It describes individual gestalts according to their significance for the spiritual contexts within which it arranges them. This *selection* is not pragmatic, to be sure, as it is in those sequence contexts that are not genuine; it is evaluative, though evaluative by the standards of unconditioned values, by the standards of validity. The pragmatic principle of selection is continued in value as the principle of selection.

But the selection in terms of value rests on a higher principle of selection. Not every spiritual act of a particular spirit-bearing gestalt is inserted within a sequence context—only those acts that constitute a historically important context. But what is the criterion for the historical importance of a context? Which acts of spiritual beings are *historical acts*? Even here one might suggest that we use the pragmatic principle and answer that for every observer, those historical contexts are important in which he himself stands. The most limited context of this kind is the observer's own history; the most extensive, the history of humanity. Between these two extremes, there is a profusion of smaller or larger circles whose development is significant for each circle itself and for everything belonging to this circle. But which individual and which circle has the historical right to make its own historical connections the principle for selecting historical contexts? The fact that a particular circle considers itself historically important does not mean that it has the right to do so.

The historical principle of selection should not be sought within the historical phenomena themselves. It can only be found in an interpretation of history that gives meaning to history as a whole. The task of the *metaphysics of history* is to provide such an interpretation from the perspective of the Unconditioned. This interpretation overcomes the pragmatic subjectivity characterizing even the individual spiritual act; it connects this act with meaning of the Unconditioned—the meaning on which all individual meaning is based and in which pragmatism and subjectivity are abolished. Whatever in history is significant for the course of this "metaphysical history" is historically important, however remote it may be from the major course of events. One need not be conscious of the metaphysics of history in every moment of historical work; indeed, such consciousness is comparatively rare. But a metaphysics of history is always present as an unconscious principle of selection; it gives the innermost significance to the construction of historical contexts. So subjective pragmatism, which governs individuals in all the empirical sciences, receives its "absolutely pragmatic" basis, its foundation in the Unconditioned, in the metaphysics of history.

c. The Categories of History

Historical causality is *productive causality*. It is causality, for every historical moment is conditioned by the totality of preceding moments; it would be impossible without them. And it is productive, because no historical moment is unequivocally determined by the past; it is free from the past. *Freedom* is the positing of the new; freedom is beginning.

The concept "freedom" is impossible within a world view defined by the method of the law sciences; when equivalent causality rules, *determinism* is always correct. *Indeterminism* is destroyed as soon as it maintains that the freedom of the will is an exceptional case within a world that is otherwise determined. Both determinism and indeterminism assume the existence of an objective reality: a world in which thought has destroyed being, in which self-contained causality (the living gestalt) is broken and the original reality (the irrational givenness) has no place. But the logical expression for pure objectivity is equivalent causality, or determinism. The concept "freedom" can only be metalogically grasped by understanding the productive causality of the original reality, the primitive state, as the prius of everything objective.

The freedom of the individual gestalt is realized in all its living acts; the original reality, the absolutely given, is present and decisive in every moment. But the limit of freedom in the gestalt sphere lies in the fact that the gestalt is the law beyond which no living act can extend. The individual gestalt realizes its freedom as creative positing of the new only where this gestalt is spirit-bearing — not in the gestalt sphere, but in the sphere of spirit. The freedom of spirit that is bound up with meaning is realized through the freedom of the individual gestalt, a freedom that is bound up with law. Spiritual creation does not destroy the forms of the individual gestalt; it breaks through them. It posits a reality of meaning, a reality that cannot be derived from any structural law, that is itself not a structural law but a unique, valid meaning. Productive causality, or freedom, in unconditionally realized in the historical creation of meaning.

Productive causality is realized in time, just as equivalent causality is realized in space. The significance of *space and time* within the three groups of the empirical sciences characterizes the mutual relations within these groups. In the physical sphere, space is dominant; in the historical sphere, time; in the organic-technical sphere, neither is dominant. Space corresponds to thought, to the fixed rational form, to the law that obliterates all individual elements and subjects the particular to itself. Space is the form of the existing "proposition," with its timeless validity. Time corresponds to being, to the infinite, irrational import, to the creative sequence that absorbs the particular into a context of meaning. Time is the form of the existing "meaning," with its creative realization. A "proposition" is as spaceless and timeless as a "meaning." But the existing "proposition" (the law) has the form of space, and the existing "meaning" (the sequence) has the form of time. The relation between the physical group and the historical group is analogous to the relation between space and time. Space is the existential form of valdity, of the immutable form;

time is the existential form of creativity, of the new revelation of import. Thus time is the form of concrete realities; it is the form of individuals. Time corresponds to the dynamic relationship between thought and being; it corresponds to the inner infinity of being, the infinity that destroys every finite form and posits new ones.

In the gestalt sciences, space and time are equally significant. The gestalt evolves in space; it is a "structure" that has achieved existence. That is directly evident in the case of the biological and sociological gestalts, but it is also indirectly true in the case of the psychological gestalt, by virtue of its necessary relationship to the two other gestalts. But space is at the same time abolished in the gestalt. The member that is indirectly evident in the psychic gestalt and that is indirectly evident in the biological and social gestalts is posited in the whole. This *space-denying spatiality* is a definite feature of the gestalt group. The gestalt is related to time in the same way. Gestalts grow, both within the species and from species to species — even technical gestalts grow out of each other. The new is posited as new, and consequently the old is posited as old; but this occurs in time. Time is denied by the gestalt law that makes every gestalt a fixed form and that makes the theory of evolution appear questionable. This *time-denying temporality* corresponds to the limited freedom and self-contained causality of the gestalt group.

Since law and sequence presuppose gestalt, time is also present in the physical sphere and space is present in the historical sphere. In the autogenous series of the physical sciences, time has become a *dimension of space*, as we have already mentioned; time is the form of development, but it is not the form of the creation of the new. In the heterogenous series of the physical group, time is the form of the development of macrocosmic and microcosmic gestalts; time reveals both the heterogenous nature of the structural character of all reality and the conquest of mere law by the infinity of existing individuals.

In the historical group, time is completely dominant. By realizing spirit, spirit-bearing individuality appears within a creative sequence of meaning of which time is the form. Here space is basically insignificant, though it is not absent. Some gestalts are indeed bearers of spirit; but the gestalt exists in space. History deals with existing and therefore localized contexts of meaning, not ideal ones; the geographical sciences are explicitly concerned with the spatial, structural foundations of history. Nevertheless, in historical investigation there is a tendency to overcome spatial coexistence and to arrive at a *universal history* that transcends space. The idea of a universal history expresses the will to overcome meaningless spatial coexistence by means of meaningful temporal succession. But that is possible only when a structure (i.e., a gestalt context) provides a foundation. Now, some social functions attempt to actualize the ideal community of mankind within a sociological gestalt. These are the functions that conquer space. To the degree that this universal sociological gestalt is actualized, and to the degree that space is abolished, according to the manner of the gestalt, space becomes insignificant for history and universal history becomes possible. Without this sociological synthesis, however, universal history remains an abstract demand; all attempts to realize it have an abstract, intellectualistic character. The universal gestalt is the prius of universal history.

d. The Methodology of History

We have already established the methodology of history in the section on the methodology of the empirical sciences. The goal of historical knowledge is the *meaningful sequence context*. It is therefore not enough to say that the individual is the object of history. The individual as such is never the object of knowledge, only the individual within a context. Nor is it enough to speak of an *individual totality*. An individual totality is a gestalt. But the historical sequence breaks through the individual gestalt. A sequence series is never, and in none of its points, a totality. It emerges from the infinite and returns to the infinite and is thus open on both ends. Of course, sequence series are based on individual gestalts; some historical concepts (e.g., "men," "peoples," "groups," "movements") refer to individual gestalts. But these objects becomes historical only when sequences of meaning, which emerge from these objects in all directions, are realized in them. Even biography is never confined to one particular gestalt. It arranges the gestalt within contexts that break through every self-contained causality, every totality. Totality is an organic category, not a historical one.

The metaphysics of organic idealism likes to apply the concept "totality" to the universe. We have seen the justification for this use of the word. The universal gestalt is a necessary idea, but the idea of a universal *cosmic history* is just as necessary. We are not referring to the heterogenous sequence series within the physical group, but to a genuine history presupposing the psychic inwardness of the cosmic gestalt. If the universe is a gestalt, it is necessarily not only an external gestalt, but also an internal one; as the foundation of all gestalts (even those that are spirit-bearing), it is itself a spirit-bearing gestalt. A universal sequence of meaning would be the prius of all particular sequences of meaning; every particular context of meaning (e.g., the universal history of mankind) would be involved in the cosmic context of meaning. But the cosmic gestalt, and therefore the cosmic sequence, is an idea. It should never be made an objective reality, as it is in myth or in bad metaphysics. The cosmic gestalt is the prius of all given contexts, but it itself is not a given context. The universal, spirit-bearing, creative gestalt is the supreme metaphysical symbol—but it is only a symbol.

The attitude of historical knowledge is perceptual understanding; its procedure is descriptive construction. The element of perception is expressed in historical research, the element of understanding, in historical empathy; description and construction are both present in historical exposition. Historical work consists of these three elements of *research*, *empathy*, and *exposition*. Each has its peculiar aim, yet they collaborate in every act of historical knowledge.

Historical research approaches its object with an attitude of objective reflection. It has developed a research technique that accords with the way facts are established in all the empirical sciences. A whole series of auxiliary sciences belongs to this aspect of historical work, including chronology, diplomacy, textual analysis, paleontology, and archaeology, not to mention ethnology and philology, which we will treat separately. These are the fundamental activities that create the material. They reveal the empirical character of all history most clearly.

The interior of historical life remains inaccessible to these activities, however. Only *empathy*, or understanding, can penetrate the depths of this life. The great achievement of the philosophy of romanticism was that it paved the way for this attitude. In the vigorous struggle against the rational, reflective attitude toward reality, the principle of "intellectual intuition" was victorious. Subject and object should not be separated; the subject must rediscover itself in its object. Thus the philosophy of romanticism created the historical sense, which produced the great achievements of nineteenth-century historiography.

Empathy leads to description. But historical description is constructive; it attempts to re-create contexts of meaning. This is expressed in *exposition*, the third element of historical work. Exposition forms a transition to the pure human sciences; therefore it occupies a peculiar intermediate position between the empirical sciences and the human sciences. This creates numerous problems. Whenever research is the dominant element, exposition is regarded as the report on the results of research. Chronographies and annals, as well as a profusion of special investigations, approximate this form. But only the external phenomena of historical life are comprehended in this way. The internal contexts, the living meaning of events, remain untouched. *Pragmatic historiography* is radically opposed to this form of history; it approaches history with alien viewpoints—political, moral, or religious—and interprets and evaluates history subjectively from these viewpoints. For this historiography, history becomes a collection of examples of moral or other ideas. History itself is not intrinsically interesting. The element of truth in this position has been appropriated by the history of spirit, which belongs within the normative sciences. But the history of spirit is possible only on the basis of history proper. In relation to history, no form of pragmatism is justified. The understanding, constructive exposition of history attempts to overcome both external objectivity and external subjectivity. It becomes the *genetic method*, which strives to re-create the internal structure of things themselves.

Sequence contexts are spiritual contexts; they become visible only when the expositor has spiritual consciousness. The limits of his creative understanding are also the limits of his re-creative understanding. The internal comprehension of every spiritual creation depends upon the *standpoint*, in the sense of the human sciences. The limits of understanding are not the limits of the standpoint itself; they are the limits of that which has positively or negatively been absorbed into the standpoint, of that which has in some way become an element of the standpoint and thus can be understood from this standpoint. This has often led to comparisons between *historical works and works of art*; universal history has been called the universal epic to which all history may contribute. The attempt is made, especially from the side of the law sciences, to assign history to the aesthetic sphere. But there is an obvious difference between the two. All history depends on the results of research. A newly discovered document can destroy the most ingenious construction. Exposition is completely dependent upon its material. This fact demonstrates the fundamentally empirical nature of history. Therefore history cannot idealize and stylize its gestalts, as the epic does; it limits itself to the given individual, with all of its existential contingencies. Historical exposition depends on the creative standpoint, to be sure, but it itself is a re-creation, a discovery, not a creation.

The degree of knowledge in history is that of *mediated conviction*. Immediate conviction is the degree of knowledge in creative construction, mediated conviction the mode of certainty in re-creative construction. Re-creation depends on the "other"; it must perceive, and insofar as it does, it achieves only probability. But as soon as re-creation moves from perception to understanding and construction, and thus as soon as the object has been re-created in the subject and made into a positing of the subject itself, probability achieves the status of conviction. This duality of probability and conviction corresponds to that of the objective and personal attitudes; it is the earmark of all great historiography.

e. Heterogenous Methods of History

In accordance with the relation between objects and methods, the historical group is also subject to the heterogenous invasion of the methods appropriate to the law and the gestalt sciences. Under the domination of mathematical physics, the scientific character of history appeared to be ensured only when *historical laws* were comprehended. There was a methodological demand for discovering the laws governing the course of history and thus for explaining the details of this process. It was thought that this is the only way to distinguish the science of history from the epic account of history. Considering the enormous complexity of historical events, this task would naturally be extremely difficult. But in principle it would be necessary; it would not be impossible.

But a closer look at the historical laws that have putatively been discovered shows that they are either *interpretations of meaning* by the metaphysics of history or laws of *sociological dynamics*. Hegel's "dialectic" is the primary example of the first class; Comte's "law of the three stages," Marx's "law of class struggle," and Spengler's "law of cultural cycles," which are grounded in social psychology, social politics, and social biology respectively, belong to the second class. But the laws for interpreting history are not laws in the sense of the physical sciences. They do not explain, they interpret. They create a systematic context of meaning; they belong to the normative sciences. The laws of sociological dynamics are genuine structural laws. As such, they are of great significance, and have an autogenous right, in sociology. They are laws of the constant forms in which social and psychic gestalts become bearers of sequence contexts, but they are not laws of the historical process itself. As soon as they attempt to become laws of the historical process, and as soon as they forget that they are heterogenous in the area of history, they destroy spirit. The use of the three aforementioned laws in the struggle against the creative conception of spirit is evidence for this assertion. Whenever the distinction between the existential form of spirit and existing spirit is overlooked, history and the human sciences become impossible.

This is also true for the heterogenous invasion of the historical sphere by the gestalt method, in the form of the *doctrine of historical structures* — a doctrine that has become increasingly influential since Dilthey. This doctrine is significant because, unlike the physical method, it recognizes the peculiarity of the psychic gestalts and has replaced explanation with description. It unites the general and the individual elements by means of the concept "the typical," in the sense of the gestalt sciences. Now, there are undoubtedly psychological and sociological types in the domain of spirit. The knowledge of these types is a heterogenous task of historical work. But this task is only heterogenous. What one can know in this way are the various kinds of containers of historical contents, but these are not themselves historical contents. They are dispositions for the realization of certain spiritual tendencies, but they are not themselves spiritual tendencies. Such dispositions are not even coercive; they can be extensively transformed by historical destiny. The spirit-bearing gestalt continually breaks through the typological limits imposed on it.

As soon as the doctrine of historical structures attempts to grasp individual phenomena, it becomes history. It tends to devote itself more to the individual structure than to the comprehensive contexts; it prefers biographies and monographs. Even biography is historiography. At times, the doctrine of historical structures contends that it is a human science. This is not so. In every one of its forms, it is an empirical science. It seeks to know existing contexts of meaning, not valid ones. The human sciences are normative.

2. The Objects of the Historical Sciences

a. Political History, Biography, and the History of Culture

Individual gestalts are the bearers of the historical process. The individual, spirit-bearing gestalt is the object of history. Thus there are two possible perspectives for pursuing historical work. One can examine the development of historical gestalts as such, and one can examine the development of spirit within historical gestalts. In the first case, one can consider either the psychic or the social gestalts. This results in three forms of history: political history, biography, and the history of culture. Political history and biography belong together as the *history of historical gestalts*; opposite these disciplines we find the history of culture, the *history of spiritual creations*. The contrast is not exclusive, of course. Even historical gestalts are spiritual creations. But they are more than that; they are also bearers of all spiritual creation. Historical gestalts are examined in political history and biography because they are bearers of spiritual creation, not because they are products of such creation. In the history of culture, on the other hand, they are treated only as spiritual creations. This is the twofold nature of history we have already mentioned in the "Foundation." But this dual nature does not imply a duality in the object. The historical material is the same in all three forms; only the orientation of the work is different.

Political history is the history of historically significant social gestalts. It is not the doctrine of sociological development, nor is it the history of social formation. The former belongs within sociology; it is concerned with constant structural laws. The latter belongs within the history of culture; it examines the development of the social functions. And political history is the history of the unique development of individual social gestalts. We call it political history for the same reason that we call the social functions the sciences of the state: because of the unconditioned priority of the power- and law-bearing community for the formation of the social organism.

Political history is not merely the exposition of political actions in the narrower sense; it is not merely the history of diplomatic acts, acts of war, and acts that create constitutions. The social organism is shaped, not only through these acts, but through all processes and functions that are significant for the development of the social gestalt. Political history thus includes all the material of biography and the history of culture. This material is not examined for its own sake, however, but for the sake of the social organism; it helps determine the development of this organism and is an expression for its actual state. But every formation of the social organism is based upon both the energy and the aim of the will that lives in this organism and that supports, protects, and develops it: the will to power and community. The formation of the social organism is based upon political acts, for the bearer of all other acts is created and transformed in these acts.

Political history is therefore not the history of politics, either in the form of the shaping of political ideas or in the form of political technique, including diplomacy, military strategy, and administration. Political history is the description of the unique effects these and all the other functions have had on the *formation of the social organism*. The decisive criterion of political history is the uniqueness, the growth and demise, of individual gestalts. Political history is therefore the natural continuation of the heterogenous sequence series. Great political movements are like astronomical and geological changes and catastrophes; they create the foundation for all culture. Yet these movements are more than just the foundation and presupposition of culture. If they were only that, they would be geographically, but not historically, important. They are, however, both the bearers and creations of culture; they are therefore historical objects, members of genuine sequence series, and spirit-bearing individual gestalts.

The methodology of idealism sees only the *spiritual side* of history. It tends either to neglect political history in favor of the history of culture or to make the former a part of the latter. This is one consequence of its neglect of the heterogenous sequence series, of which political history is the continuation. Idealism overlooks the existential, gestalt-like nature of all history. But whenever the individual gestalt is seen as the bearer of history, the dual nature of all history becomes clear, and political history regains the significance it has always had. Of course, it should not be merely the report of battles and dynasties, as it has often been (partly under the influence of biographical and epic tendencies), provoking the justified opposition of the history of culture. Political history is the history of individual social gestalts: that is its meaning and its fundamental significance for all other forms of historical work.

What is true of political history in relation to social gestalts is also true of *biography* in relation to psychic gestalts. Biography is the history of historically important members of the social organism. It is therefore based on political history, yet it is not merely a collection of monographs on political history. It is concerned with particular individual gestalts, whose development is indeed interwoven with the total development, but who also have a special existential roots, a distinct gestalt, and a special fate. Biography is also related to the heterogenous sequence series. This relationship becomes especially clear when one traces biography back to its epic and mythic origins. Divine acts and cosmic fates are the primary contents of sacred histories. But divine actions revolutionize heaven and earth; they are cosmic catastrophes that determine the present form of the world. Even heroic acts change the face of the earth. Thus in the beginning, *natural events and history*, heterogenous and autogenous sequence series, are closely connected. Then gradually the side of personal heroism predominates, and the heroic epic transforms heroes into spiritual individuals. But the irrational elements of power, passion, and fate still dominate. Only later do the spiritual creations become significant, though they are unable to suppress the irrational elements in genuine biography. The onesidedness of the methodology of idealism is most obvious in biography, which is the oldest and most original form of the comprehension of historical reality.

Political history and biography are filled with cultural contents. The *history of culture* is the investigation of these contents. It is not concerned with spirit-bearing gestalts, but with spiritual contexts. Every spiritual and technical function, from the history of a craft to the history of religion, can be the object of the history of culture. Political and geographical material is always present in these histories, but this material is not the goal of knowledge. The goal is the development of the spiritual functions, or spiritual sequence contexts.

Existing contexts of meaning can be considered either from the point of view of their existence or from that of their meaning. The former is the perspective of the history of culture; the latter, of the *history of spirit*. The history of spirit does not belong within the historical group, however. It is indifferent to the existence of contexts of meaning; its task is the creative understanding of meaning. The history of spirit is a genuine normative science and a basic element in all the human sciences. On the other hand, the existence of contexts of meaning is important for the history of culture. This discipline seeks to know what spiritual creations there are, how they are constituted, how they are genetically related, what influences they exert, and so forth. The history of culture views spirit within the sphere of causality; the history of spirit considers it within the sphere of validity. Naturally, no work in the history of culture is possible apart from an understanding of the contexts of meaning. The history of spirit is present in every presentation of the history of culture that is more than the mere communication of the results of research, and every history of spirit contains material from the history of culture. But the orientation in each case is fundamentally different: in the one case, it is toward the existence of the meaning, in the other, toward the meaning of the existing.

Every realization of culture contains a technical element. Therefore the history of culture is immediately related to the *history of technology*; indeed, the history of the technology of development can be presented from either a technical, heterogenous perspective or from a cultural, autogenous perspective. One can show either the realization of goals or the relationships among goals. Like all history, the history of culture is related to the heterogenous sequence series; it leads these series from the sphere of being into the sphere of meaning.

Political history, biography, and the history of culture are the three basic standpoints of historiography. In each of these disciplines, the standpoint of the historian is significant — whether it be his membership in a social organism (in political history), the standpoint of the history of culture (in the history of culture), or the limits of individual understanding (in biography). But the pragmatic subjectivity of selection is abolished, in principle, within metaphysical history and its interpretation of meaning from the perspective of the Unconditioned; the subjectivity of understanding is overcome to the degree that the historian succeeds in going beyond every individual, sociological, or psychological standpoint to the *absolute standpoint*, which interprets and understands from the perspective of the Unconditioned. But this absolute standpoint, which intends to abolish the standpoint as such, lies beyond the contrast between autogenous and heterogenous history. The absolute standpoint is the interpretation of the meaning of the cosmic process as such. But it is no longer science; it is metaphysics.

b. Anthropology and Ethnography

We turn now to some sciences that are related to the historical group but also extend back to the organic group. The terms "anthropology" and "ethnography," "linguistics" and "philology" refer to complex sciences that center around an object belonging to various groups, both methodologically and as far as objects are concerned. The task of systematics here is to dissolve these *complex sciences* and to divide their elements into systematically independent groups. We have frequently practiced this procedure already; but in the present cases, the situation is somewhat more difficult, requiring a special discussion.

As the object of *anthropology*, man is first of all a biological gestalt. Therefore, the first task of anthropology is to apply the biological laws of structure and evolution to man and to determine his generic characteristics. The generic structure of man is individualized in the human races and their subspecies; the investigation of the structural forms and structural laws of these races and subspecies is the further task of anthropology.

Like all living beings, man is also a phenomenon appearing on the surface of the earth. He is thus an object of *geographical anthropology*. One cannot state the relationship between man and the earth's surface unless one considers both the influence of man on this surface and the ways in which he subdues it. Then anthropology merges into the history of technology. But this is not feasible apart from the inclusion of psychological and sociological elements. As soon as psychic and social contents are considered, however, anthropology becomes history. Thus anthropology has its special area only in biology.

Anthropology has certainly not analyzed this situation. It has continually invaded the other areas. The reason for this is that from the standpoint of political and cultural historiography, a large portion of mankind is usually considered *prehistorical* or *unhistorical*. Archaic man is called prehistorical because he has produced no extant historical documents; primitives are called unhistorical because they have produced neither states, properly speaking, nor individual culture. So prehistorical and unhistorical mankind are left to anthropology and ethnography. This attitude is determined by the standpoint of the law sciences, a standpoint that believes it is best able to establish typical structures and universal laws of development in relatively unformed humanity. Thus anthropology becomes the description of that portion of mankind existing outside history—that portion in which there are said to be constant forms but not genuine sequence contexts.

This position is indefensible, systematically speaking. The concept "prehistory" is based upon the criterion of literacy. There are not enough documents from the most archaic periods, but there are documents; from these documents it is even possible to discern genuine sequence contexts. The idea of a primitive unhistorical mankind is just as questionable. Whenever one penetrates the essence of *primitive cultures*, one finds cultural and social creations of individuality and (some) excellence. Although the typical stands out more strongly here than in the great cultures, the individual is still significant enough to be arranged within genuine sequence contexts. Anthropology is heterogenous in the historical sphere, from the very beginning of human evolution.

Ethnography is the description of racial conditions, considered apart from the element of sequence. Thus its objects are those phenomena within the life of races in which the permanent prevails over the changing. Ethnography assimilates anthropology but extends it by investigating racial individualities. It is the description of racial gestalts. But it cannot ignore spiritual individuals. Ethnography merges into history. In fact, it has even assumed the function of collecting material for the use of history; therefore, it belongs to the auxiliary sciences of historical research.

c. Linguistics and Philology

Linguistics and philology are complex sciences. Their most general object is language. Language is first a psychophysical phenomenon; thus it is the object of *the physiology and psychology of language*. But language is also the form for the direct expression of spirit; as such, it has two aspects: a universal aspect, by which it is related to logic, and a particular one, by which it expresses a concrete spiritual situation. In the first respect, language contains the universal form of thought, logic. The task of the *logic of language* is derived from this. This discipline is not concerned with logical propositions and structures as such, as is pure logic; it is concerned with the way these propositions and structures are verbally expressed. It is therefore an empirical science, not a science of thought, although by the influences it exerts upon logic it can also be quite significant for the pure sciences of thought. But the logic of language belongs within cultural psychology's theory of structures. *Grammar* is the introduction to the construction of individual languages. Comparative grammar especially is the material foundation of the logic of language. Grammar itself is not logic, however, though it presupposes logic and has been the great instructor of logical thought. It is a historical, individualizing empirical science. The laws of grammar are descriptions of the individual structure of specific languages rather than of the laws of language in general. Only the logic of language is concerned with the general laws of language.

Semantics is naturally more interested in describing individuals than is grammar. Semantics contains biological and psychological elements and thus belongs in the sphere of the description of individual, heterogenous gestalts. But that is only one aspect of this discipline. The way in which a language-creating community spiritually grasps reality is always clear from the way it forms and transforms words. And in this respect, words are objects of historical investigation.

Grammar and semantics are placed within more comprehensive contexts by *comparative linguistics*. This discipline discovers both the origins of speech and the relationships among the individual languages. As far as its goal is concerned, comparative linguistics has aspects of the sciences of both history and law; it seeks to know the history of language and the laws of linguistic development. Actually, both are possible because of the dual nature of language: as the expression of spirit, the development of language is historical; as the act of expression, it is psychophysical. The result is the peculiar situation that both the historical method and the method of structural law are, when applied to the evolution of language, always both autogenous and heterogenous. This corresponds exactly to the duality we have identified in all the historical sciences.

Language, which is the expression of spiritual contents, has created an expression for itself in *writing*. Writing is more technical than language, because of its greater distance from spirit. The investigation of writing belongs within the history of technology; it can be evaluated by autogenous history only on the basis of the most general presuppositions.

Finally, language is material for artistic creation. In this

role, it becomes an important object of philological investigation. Thus philology encroaches upon the aesthetic domain; by
the exposition of the *history of linguistic style and of literary
forms*, it performs a task proper to the history of culture. But
this is only possible because philology itself attempts to grasp
the content expressed in the spoken word and in writing. In
this way, philology becomes an element of historical science
and indeed of all areas where spirit is expressed in the word. It
becomes *interpretation*, the understanding of past realizations
of meaning. But philology itself is not history; it creates
material and shows how to understand it. As soon as it becomes
exposition, it is history, not philology.

In summary, we can say that linguistics and philology constitute the complex of those sciences that both deal with the
word as an independent reality and seek to understand
historical reality from the perspective of the word.

PART 3

The Sciences of Spirit, or Human Sciences (The Normative Sciences)

I. Foundation

A. *Spirit*

1. *Spirit, Thought, and Being*

The third group of sciences is composed of the "sciences of spirit," the human sciences, which are the normative sciences. The *concept "spirit"* is not as fundamental as the concepts "being" and "thought," for it is dependent on them: spirit is the self-determination of thought within being. It is impossible to grasp the essence of spirit without metalogically grasping the two basic elements of knowledge. The essence of spirit, its inner tension, its dynamic character, is derived from the infinite opposition between thought and being. Logism's view of spirit usually neglects the element of being; psychologism's view neglects the element of thought; both neglect the tension between the two elements. But spirit is neither a mode of thought nor a mode of being. In spite of its dependence upon both of these elements, it is an irreducible mode. Spirit is the mode of existing thought.

Every existent is formed by thought, and the more thought being has absorbed, the more real the existent is. The being with the most reality is the individual spirit-bearing gestalt, that being in which thought is realized as thought, as valid form. In all other being, thought is realized as a form of being, as conditioned, limited, immediate form. In the spirit-bearing gestalt, however, thought escapes from its conditionedness and immediacy; it confronts all forms of being with the unconditioned nature of its demand; it confronts being as *validity*. The presupposition of the realization of spirit is therefore the complete separation of an existent from the immediate subjection to its finite form. The presupposition of spirit is freedom. There is an unconditioned demand only for the free individual gestalt; validity can be realized only on the basis of freedom. This is the absolute boundary separating spirit from the psychological and sociological domains. Spirit is first actualized when validities, which are subject to their own law of meaning, are grasped in individual gestalts, which are subject to their own structural laws. Therefore, every spiritual act breaks through the limits of the immediate gestalt — it breaks through these limits, it does not destroy them. For it is the nature of the spirit-bearing gestalt to realize within itself something that does not derive from it: validity.

The thought sciences are also concerned with validity. Their structures and propositions express the eternally valid form of pure, self-sufficient thought. This is not yet spirit, though the comprehension of thought is a spiritual activity. Spirit first appears when existents are subjected to validity, when the unconditioned demand is absorbed into being. This is the *bond between spirit and being* that logism overlooks, just as psychologism overlooks the freedom of spirit from being.

The functions of the spirit-bearing gestalts are subject to unconditioned validity. On the one hand, the spirit-bearing gestalt is completely separated from the universal; it is something absolutely unique and individual. On the other hand, it contains the universal; it can absorb everything real. But everything it absorbs receives the peculiar formation corresponding exactly to its individuality. In the face of reality, every gestalt is both receptive and reactive, absorbing and formative; both of these aspects constitute one act. This act is spiritual insofar as it is determined, not by the immediate life of the gestalt, but by the unconditioned demand, by validity. All psychic and social functions in which every act of formative absorption is realized — that is, all functions in which the freedom of the spirit-bearing gestalt is represented — are subject to spiritual formation. The task of the system of the human sciences is to designate and organize these functions.

Spirit receives its *concreteness and fullness*, its individuality and infinity, through its relationship to the functions of the spirit-bearing gestalts. This is what distinguishes spirit from the pure forms of the thought sciences. Nevertheless, in all the concreteness and particularity of the spiritual act, there is still the universal, the valid, the unconditioned demand. The fundamental category of spirit, creativity, depends on the indissoluble connection between this universality and the particularity of the individual gestalt.

2. Creativity

Creation is *original positing*. Thus all derivative positing — everything that can be deduced, explained, reduced to something else — stands in contrast to creation.

Every immediate relation to the world is expressed in concepts that take into account the creative character of being. Cosmogonic myths are symbols for this. Even the concept "creation" in dogmatic theology is a rational vestige of the primordial conception of the world. But as soon as rational thought objectifies the world and things, creativity disappears and is replaced by equivalent causality, law, and evolution. Everything becomes objective, nothing remains primordial. Of course this attempt is not entirely successful, for at some point the original givenness of being must be acknowledged; but this point is pushed back to a problematic beginning and plays no role in comprehending the world. The victory of the method of mathematical physics in all areas of knowledge has eliminated the concept "creativity" from the understanding of the world. Of course, this concept had a secondary influence in the biological controversy over the spontaneous generation of life, but it could not maintain itself within the law sciences. The doctrine of individual gestalts is a conscious return to the *creative, primordial conception of the world*: every gestalt is an original positing. And the metalogical understanding of the category of substance provides the epistemological foundation for this conception.

Every creation contains two elements: an element of being, through which creation becomes original positing, and an element of thought, through which creation becomes determined, formed positing. For being is the principle of original positing, of unconditioned reality, and thought is the principle of form, of unconditioned validity.

Every creation unites thought and being, so every creation is both individual and universal. Creation is the *individual realization of the universal*. The more individual and at the same time more universal a reality, the clearer its creative character. The highest form of creativity is thus the spirit-bearing gestalt. Equivalent causality, with its law and necessity, overlooks creativity; productive causality, with its individuality and freedom, expresses creativity completely. The spiritual individual is creativity, in the true sense. Religious symbolism has found an expression for this relation by letting man participate in the divine creative potency and by anticipating the completion of the divine work through human work. The abyss of the Unconditioned is revealed more profoundly in man's spiritual creations than in the gestalts of immediate existence, which are always subject to law.

3. Spiritual Creation

Like all creation, spiritual creation contains both a universal and a particular element. It is not as if these two elements were first separated and were then united; rather, only the *creative unity* is given. The fact that this union contains two elements is a later abstraction. It is therefore impossible to grasp the two elements separately. Whoever attempts to do this will have empty matter in one hand and empty form in the other, but he will have lost the spiritual reality. It is not difficult to recognize this in the case of the latent creations of the gestalt sphere, where the unity of the universal and the particular is immediately given. But it is harder to recognize this in the case of free creations of the spirit, where the realization of the universal in the particular occurs by means of conscious activity, which in the spiritual process can be directed only to the universal.

This situation has led to two opposing interpretations of spiritual creation. One emphasizes the rational element; the other, the irrational element. The rational interpretation regards rational creation as the attempt to realize the universal: the more the creation succeeds in this, the more valuable the work is. The universal is the law to which the individual must submit; insofar as this individual does not do so, it is deficient and incorrect. All spiritual creation is an *approximation to an infinite ideal* that is more or less approachable. Particular works are to be judged by this quantitative criterion of approximation. This conception is undoubtedly determined by the quantitative method of law, in which individual creativity has no place. For a philosophy whose categories are oriented to mathematical physics, spirit can only appear as the attempt to realize a law.

The other interpretation relinquishes the universal, seeing spiritual creativity in the immediate development of the individual gestalt. That which breaks through the depths of the biological and psychological dimensions is creative positing; indeed, it does so in inverse proportion to the degree that it is influenced by the universal. Therefore, whether one emphasizes *subjective feeling* or *subjective will* (in the sense of pragmatism), the mark of creativity is always the denial of the universal element. But this position forgets that in the gestalt sphere, no perfect individualization is ever attained, that structural law rules and arbitrariness is nothing but biological or psychological necessity. Only where the universal is present in the particular is there real freedom and true creativity. For whoever wishes to escape the law of meaning is subject to the law of being.

The creative act therefore contains both the intention toward the universal and the character of individual gestalt, of existential matter, in which the universal is realized. The truth of the rational view is that the intention can be directed only toward the universal. If the intention were directed to the individually given, it would not aim to realize spirit and validity; the individual would cease to be material. Instead of developing its creative powers, the individual would lose its immediacy and be destroyed. The will to be individual destroys the creative power. Yet the intention to the universal can be realized only in the individual gestalt. Realization means becoming one with the existential individual. If the universal would remain universal, then it would remain an abstraction and would not be realized. Creation and spirit are nothing but the unity of *intention toward the universal and realization within the particular*. Here, too, religious language has created a symbol: in contrast to both lawfulness and lawlessness, it proclaimed blessedness the state of being filled with the divine. Spiritual creation is "grace."

Spiritual creativity is not the same as genius, however. Every act of spiritual realization is creative and original. No spirit-bearing gestalt is excluded from creation. The difference is only that most individuals are original not by virtue of their individuality, but by virtue of their being members of a social organism. Their creativity is part of the creativity of the whole. But the creativity of the social gestalt finds its highest expression in the creations of a few outstanding individuals in whose individual creations the spiritual potentialities of the entire social gestalt (under certain circumstances, all of humanity) are actualized. In these individuals, spirit becomes *genius*, or blessedness in the transcendent sense. They even break through genius to a second nature, to a kind of gestalt law for current forms of community. Yet they are dependent on the social gestalt out of whose substance they create. They can no doubt break through this gestalt, but they cannot destroy it. Even the highest genius is bound to the material of the individual gestalt; every social gestalt — even humanity — is only a standpoint, an individual material for the realization of spirit. Even the human spirit is not the universal. For spirit is infinite, and it becomes real only in individuals, whether in a particular individual, in humanity, or in a comprehensive spiritual realm.

4. The Limits of Creativity

The rationalistic position can criticize the individualistic conception of creativity by claiming that this conception is itself a creation of the human sciences and thus cannot claim to be universally valid. This is the objection that is rightfully raised against every dogmatic skepticism and dogmatic *relativism*. It does not apply to our own conception, however. This objection presupposes that to be individual is to be incorrect. But we dispute this presupposition. Truth is not a system of abstract validities in relation to which knowledge of the truth is either correct or incorrect. Truth is a function that is realized only concretely and that is realized correctly in every creation. In the realm of spirit, the antithesis is not between correct and incorrect but between creative and arbitrary. Whatever is arbitrary in a position succumbs to the judgment of the spiritual process, history; whatever is creative remains a living moment in the entire spiritual process. Both the judgment concerning arbitariness and the justification of meaningful necessities arise in the struggle of convictions. But the judgment does not occur in such a way that a correct and an incorrect element are separated; it occurs in such a way that a new creation posits itself against the old one and absorbs it. The driving force of the spiritual process is not abstract criticism but the positing of the new. Criticism is only a function of creation, just as law is only a function of gestalt. When it is separated from creativity, criticism is empty and futile. Spiritual creation is the prius of all spiritual criticism, not conversely.

This discussion has indicated one of the limits of creativity: the universal. Whatever lies beyond this limit is arbitrariness. Opposite this limit of principle stands an empirical limit: the limitation of the individual gestalt and its creative substance. Spirit is inexhaustible, for being is inexhaustible. No individual realization of spirit could exhaust spirit; there is no absolute standpoint. But the individual gestalt is not inexhaustible. For every individual gestalt, there are contexts of meaning whose assimilation would not only break through its gestalt structure, but would also destroy it. The finite gestalt cannot realize infinite sequences of meaning within itself, although as far as meaning is concerned every context of meaning has infinite relations. Every individual spirit-bearing gestalt is therefore subject to the *law of the exhaustion of substance*. The exhaustion of spiritual substance is not identical with the exhaustion of organic substance, or death; one can occur without the other. This mutual independence is not perfect, however. Because the nature of the spirit-bearing gestalt is to realize spirit, spiritual exhaustion eventually leads to organic death. This is true not only for the individual gestalt, but also for the social gestalt, as well as for all humanity and for every conceivable spirit-bearing individual gestalt.

The creative power of all of them is limited by the fate of the exhaustion of substance.

It follows that the successive spiritual realization in an individual gestalt represents neither an infinite progress nor an unlimited alternation between progress and regression, nor an uninterrupted series of equally valuable creations. They represent a limited number of creative breakthroughs, which, in their totality, exhaust the individual substance of a gestalt. The fact that such an exhaustion of substance occurs has no significance for the spiritual act itself. There is no theoretical or practical realization of spirit in which the consciousness of exhaustion would be a constitutive element. Any attempt (by the philosophers of decadence and decline, for example) to include this consciousness in the spiritual act would simply prove the inexhaustibility of substance. Exhaustibility is an empirical limit, not a limit of principle; it is actually, not intentionally, contained in spiritual acts. The spiritual act itself knows only one limit: the unconditioned form, or validity.

5. Spirit and History

The relationship of creativity to history follows from the union of its two limits. The individual substance of the spirit-bearing gestalt is not unformed chaos, for the spirit-bearing gestalt is always a spirit-formed gestalt as well; this gestalt stands within a historical sequence series. It creates from within its individual historical formation. This formation reaches back beyond its own gestalt origin, through the comprehensive context of all gestalts, to the universal and infinite gestalt, which is an idea, not a reality. A historical gestalt has no moment outside history. There is *no origin of spirit*, for every spiritual creation presupposes spirit.

The spiritual act contains the intention toward the universal. But this intention would necessarily have remained without content if it had sought to grasp the universal as universal. For the unconditioned form does not exist as a reality that can be grasped. The spiritual act can be directed to the universal only when it intuits the universal in a concrete norm, in an individual realization of the universal. Therefore, all spiritual creation is oriented toward history. The most powerful spiritual creations in theonomous and autonomous culture are most emphatically directed to the past. Revelational authorities, classical periods, romantic attempts to return to the past—these phenomena signify the consciousness that all spiritual creation depends on the concrete realization of spirit, on history. Only an uncreative belief in progress is immediately directed to the universal. It wants to realize the universal abstractly. But this belief does not realize the universal at all, or it realizes the universal only insofar as it is itself still dependent upon history and is fulfilled with concrete relations of meaning. For orientation toward history does not mean objective historical investigation; it means fulfillment with the concrete wealth of the past. This can occur in the least objective knowledge of history, and it can fail to occur in the most exemplary historical knowledge. *Historical consciousness* is being filled with the forces and tensions of the historical process; it is being aware of the creative significance of the present moment.

It follows that to turn toward the past is not to be bound to the individual element in past creations. No individual gestalt can submit itself to the individual element in another gestalt. If it attempts to do so, it is destroyed. Such destruction is a characteristic of the *heteronomous approach* to history, which is frequently found in certain religious authorities, in contradiction to the nature of creative spirit. This approach to history is on the same level as being directed toward the individual element of one's own gestalt and is frequently a reaction against this. In the spiritual realm, however, creative intuition has only one direction: the direction toward the universal. Every deviation from this direction, whether it take the form of being directed to one's own or to a foreign individuality, is disobedience to the unconditioned demand of validity; it is destruction of the creative act. Of course the act of creation can realize itself only in the individual gestalt formed by history, whether this gestalt be the individual, a group, or mankind. Both determination by history and freedom from history are crucial for the creative spiritual process.

B. *The Human Sciences*

1. *The Productive Character of the Human Sciences*

Like every other spiritual act, *science* is creative. Thus every act of knowledge contains both a universal element and an in-dividual element. Naturally, the universal predominates to the extent that the object of knowledge itself approximates the universal, the pure form of thought; thus the universal is most clearly prevalent in the thought sciences. But the individual predominates where the object is least capable of being grasped by the determinations by thought—that is, in the human sciences. The relation between the individual and the universal in the sphere of science is one of the most difficult and most important problems in the theory of science. We will consider this problem in the "Conclusion."

The peculiarity of the human sciences lies in the peculiar relation they have to their objects. Unlike the other sciences, the human sciences never deal with given objects that they can examine and assimilate, whether by intuition, perception, or empathy. They always participate in the positing of the objects they seek to know. They are not only re-creative, like history; they are also co-creative, or *productive*. This is because every act of spiritual creation combines the individual substance with an act of consciousness that is directed toward the universal. For it is the gestalt that is completely free, by virtue of its separation from all immediate connections, and is thus com-pletely conscious—it is this gestalt that becomes the spirit-bearing gestalt, because it is directed to the universal. The fun-damental characteristic of spirit is this consciousness, this self-observation and self-determination of thought in the creative act. One co-creative element in every spiritual act is con-sciousness, or being directed to the universal and the valid. This is not the only element, for creative substance, the living gestalt with its immediate existential relations, operates alongside or within it. The spiritual act requires the coopera-tion of both elements. But consciousness, or being directed to the universal and the valid, is part of every spiritual creation; it is a productive element of the creative process.

Consciousness itself is not yet science, but it can become science; it can develop from the accidental, the instinctive, and the artificial to clarification of fundamentals and systematic understanding. The *consciousness peculiar to the science of spirit* can develop from the consciousness peculiar to spirit.

Not every proposition in the human sciences is productive in a direct way. Within the sphere of science, only that which pertains to the scientific form is creative in the highest sense. But directedness toward the whole, the norms established in a system of the human sciences, can be crucial for the intentions of spiritual creation. The conscious side of the creative process can be completely determined by these norms. In the aesthetic, social, religious, and other spheres, this relationship can be shown by innumerable historical phenomena. This is also true in the sphere of knowledge, for even in science there is both work that is instinctive and artificial and work that is methodological and conscious of fundamentals. But when a system of the human sciences operates productively in this way, it does not create the universal it formulates; rather, the system collaborates in individual creations, which then present themselves to it as new objects of systematic understanding and lead to formulations of the universal that are new and creative. The *productive character of the human sciences* does not entail a rationalization of the spiritual process. This would happen only if it were possible to avoid the fact that all spiritual realization is individual and existential—the fact neglected by logism. Spirit is never merely universal, however; it is always universal as the formation of an individual gestalt.

Alogism is just as wrong when it disregards the element of consciousness, basing the creative process entirely upon instinct and feeling, making the human sciences another empirical investigation of the spiritual process, making these disciplines history or typology. Whenever there is no intention toward the universal, arbitrariness replaces creation. The human sciences are productive; that is, they are always both the prius and the posterius of spiritual creation. They live from the creations they co-create; they help posit the objects they know. They are the systematic form of self-determining thought.

2. The Normative Character of the Human Sciences

The normative character of the human sciences is immediately given with their productive character. Norms are laws for spiritual creation. The human sciences become lawgiving through the knowledge of norms. The doctrine of science gives laws for scientific work; the doctrine of art gives laws for aesthetic intuition; the doctrines of community and law give laws for social and legal action; metaphysics, ethics, and theology give laws for the theoretical and practical, the autonomous and theonomous, attitudes toward the Unconditioned. The human sciences are *normative sciences*.

This statement is insufficient, however. It could be interpreted to mean that the knowledge of norms is an independent act like those within the ideal or empirical sciences — an act similar to the knowledge of logical laws or anatomical structures, for example — and that the laws known in this way would then have normative significance for spiritual creation in a secondary way. The possibility of using these laws normatively would not change their content. For this reason, it would also be inappropriate to speak of normative sciences, because the normative element would not be their distinguishing characteristic.

This view presupposes the logistic interpretation of the spiritual process. On the one side, it sees the rational norms, which are more or less discernible; on the other side, the activity that is directed to their realization. Of course, it does not matter whether the norms are realized in activity. Their validity is independent of their realization. But the theory of the creative nature of spirit denies this logistic presupposition. Norms are born in the creative spiritual process. They have reality only through this process. They do not exist in an ideal sphere, as do the pure forms of the thought sciences; nor do they have immediate reality, as do the structures of the empirical sciences. They have the peculiar reality that the creative spiritual process gives them. They enter this process, emerge from it newly formed, and reenter it. They originate from spiritual creation, and the latter is directed to them. They are individual when they are posited by the creative process; they represent the universal when creation is directed to them. The universal itself does not exist, however. All spiritual forms are directed to the *unconditionality of the pure form*, but the Unconditioned itself is not a form. The Unconditioned is the Eros of spirit, the Eros that turns toward spirit; it is the longing of all conditioned forms for the Unconditioned. It is not obedience to a law, to a system of concrete norms that are more or less known, more or less realized. The normative character is thus not something that is added to spiritual creations; it is something essential to them. This normative character reveals that the nature of spiritual creations, their existence as spirit, is the self-determination of thought within being. Every spiritual act is the positing of a norm; insofar as the human sciences productively participate in this act, they are normative. The normative character of the human sciences consists in the fact that they help create the norms that they know.

Today the human sciences are often called *sciences of value*. The concept "value" refers to the subject that the spiritual confronts with the claim of validity. But this relation is conceived from the point of view of the subject. Spirit is primarily validity, truth, norm; only secondarily is it value. It is not appropriate to define spirit in terms of a secondary characteristic, especially since this characteristic is not sufficiently defined by the concept "value." There are numerous values (e.g., biological, psychic, and economic values). But spiritual values are unconditioned values. Without the qualification "unconditioned," the definition of spirit as value is misleading. The problem of values belongs within ethics. But "value" is not a comprehensive concept that could apply to the entire domain of spirit; it includes both too much and too little, and it contains an element of subjectivity that is avoided by the concept "the human sciences," or "the normative sciences."[1]

3. The Goal of Knowledge in the Human Sciences

The cognitive goal of the human sciences is *meaning*. The context of meaning in the human sciences corresponds to structure in the thought sciences and gestalt in the empirical sciences. In every moment of its existence, each gestalt performs acts relating it to the rest of reality: acts in which it absorbs things, acts in which it penetrates things. This dual act constitutes the life of the individual gestalt. In prespiritual gestalts, this act occurs in an immediate way; it is subject to structural laws. In the spirit-bearing gestalts, the act comes under the demand of the unconditioned form. The relationships are raised from the sphere of being to the sphere of meaning. The acts of the spirit-bearing gestalt are meaning-giving acts. This is not to say that a reality that has no meaning becomes meaningful through the acts of spirit-bearing gestalts. Such a pragmatic interpretation misunderstands the nature of spirit. Rather, the meaning-giving acts are meaning-fulfilling acts. The meaning inherent in all existential forms comes to itself in spiritual acts; the meaning of reality is realized in spirit. All being is subject to the law of the unconditioned form, but the Unconditioned is grasped as Unconditioned, as validity, only in spirit. The meaning of being is fulfilled in spirit.

Like structure in the thought sciences, the context of meaning exists in the ideal sphere; it has the character of validity. Therefore, one also speaks of "structures of meaning," combining in this concept the cognitive goals of both the thought and the human sciences. But this use of language is inappropriate, because it erases the fundamental distinction between the two kinds of science. The structure found in the thought sciences is a pure form, which is independent of the spiritual act that is directed to this form. To be sure, the structure can be made the object of a fulfillment of meaning, but that is not its nature: the pure forms of thought are conditions, not creations of a meaning-giving act. On the other hand, spiritual contexts of meaning do not have an abstract existence. They are realized in the concrete fullness of phenomena. They continually bring new objects to meaning-fulfillment. They live in the acts that realize meaning. A context of meaning is not a structure; it is a unity of meaning-fulfilling acts. We call this unity, which is grasped by a concept, a "*system*." A system is distinguished from a structure by the fact that the system is a unique and creative comprehension of meaning. It is not a thought form about which one can make propositions; it is not an existential form about which one can assert laws. It is a positing that has no other existential ground than the spiritual act by which it is posited. The propositions of the human sciences are not propositions *about* systems; they are propositions *within* systems, propositions in which the context of meaning is represented. This is true even when the systematic form is not achieved. In intention, every statement made by the human sciences is directed toward the system; in fact, there is no statement that is not related to a basic systematic position. It does not matter whether the system exists in reality or only ideally. Methodologically, all work in the human sciences consists in the formation of a system. Meaning is system.

The systematic character of the human sciences is based on their normative character. To establish norms is to place under the unity of a valid principle. To arrange something within a context of meaning is to subject it to the unity of the unconditioned form. The sciences of thought do not require a system, because the unity of the thought form does not confront any real diversity of objects; the empirical sciences do not need one, because the infinity of being eludes unity. To be sure, it is possible to present the thought sciences and the empirical sciences according to the analogy of the system; but a system within these kinds of science is always rendered obsolete by new findings and disclosures. Such a system is a summary from a uniform point of view, but it is not a genuine system. By contrast, the system within the human sciences is closed; it is a *unique, creative positing* that can be replaced but not violated. The genuine system is an individual creation. It is distinguished from spurious, heterogenous systems in the same way that genuine gestalts are distinguished from heterogenous gestalts that are disregarded by the physical process. Of course it must be noted that the creative nature of science means that the methodology of the human sciences influences the other areas. Whenever these influences are felt, whenever the creative character of science is effective in the different areas, we find genuine systematics: for example, in Euclidean geometry as distinguished from non-Euclidean, in classical mechanics as distinguished from modern, in materialistic sociology as distinguished from idealistic. But the systematic element in these areas is based on the standpoint of scientific theory, not on the object. The self-contained system is the autogenous goal of knowledge only in the human sciences.

The system of meaning is based on the tension between two elements: the principle and the material of meaning-giving. When one examines these elements, the system appears as the third, synthetic element of the human sciences. The work of the human sciences is accomplished in three parts: the *doctrine of the principles of meaning*, the *doctrine of the material of meaning*, and the *doctrine of the system of meaning*. To a certain extent, each of these three areas can be treated separately, though systematics, which is dependent on the other two, is least receptive to separate treatment. But the complete science of spirit is the union of the three within the normative system. In "The System of the Human Sciences," we will examine the doctrine of the principles of meaning under the heading "philosophy," understanding by this term the doctrine of the spiritual functions and categories. The doctrine of the material of meaning is identical with the history of spirit, a discipline we encountered in the discussion of the historical sciences. We regard this discipline as the second element of the human sciences, in addition to the first element, philosophy. Following these is the third element, systematics, in which the mutual relations among the three elements are clarified.

4. The Attitude and Procedure of Knowledge in the Human Sciences

The cognitive attitude of the human sciences is that of *understanding*. Contexts of meaning are understood. Understanding contains an element of rational intuition as well as an element of rational perception, or experience. But both elements are completely transformed in the human sciences. Understanding is neither intuition of pure forms nor experience of strange being; it is conscious participation in the creative act of meaning-fulfillment. Understanding is the consciousness accompanying the spiritual act itself, raised to systematic methodology. Understanding in the human sciences is productive; this distinguishes it from historical understanding, which is re-creative and empathetic.

Understanding is different, however, in each of the three elements of the human sciences. *Productive understanding* is the peculiar characteristic of systematics. In systematics, where the act of positing norms determines all contexts, the productive character is strongest. Productivity is also present in the other elements, but in them it has connotations of both the thought sciences and the empirical sciences. The history of spirit receives the material of the concrete realizations of meaning that history presents to it. It presupposes historical, re-creative understanding. Its understanding is also productive, for it is guided by the norm. But unlike the pure productivity of systematics, the history of spirit is still partly determined by the material. It is not just productive understanding; it is also *receptive understanding*.

Philosophy extracts the principles of meaning-fulfillment from the concrete fulfillments of meaning. It separates the relatively constant elements from the variable elements in the giving of spiritual meaning. Philosophical understanding is therefore *critical understanding*. It is not that the productive element is missing. Even the functions and categories are not pure forms, like the ideal structures; they, too, depend on the process of normative meaning-giving. But their kind of dependence is different from that found in systematics and the history of spirit.

The procedure of knowledge in the human sciences, *construction*, corresponds to the system as the goal of knowledge. In scientific discussion, the word "construction" has received connotations of both the mechanical and the fantastic. Both connotations deviate from the real meaning of construction, the former in the direction of abstract schematism, the latter in the direction of unfounded speculation. Systematic construction is continually threatened by both dangers, but its nature remains unaffected by these threats. Construction is the presentation of contexts of meaning from the perspective of a normative principle. As such, it is the legitimate form of knowledge within the human sciences; it is even used by those who resist it. For it is necessary to use construction; its use is grounded in the nature of spirit itself. Construction becomes fully effective in systematics, where it is *synthetic construction*. In the history of spirit, it places the material within normative contexts. Construction assigns particular phenomena their proper place within the total context; it should thus be called *arranging construction*. In philosophy, *analytical construction* corresponds to critical understanding; both of them are concerned with polarities. Analytical construction is construction, for it understands the contexts within the areas of meaning and categories from the perspective of the ultimate norm of meaning-giving; it is analysis, for it constructs by separating the principles of meaning from the contexts of meaning.

The kind of certainty found in the human sciences is *convic-*

tion, which is an original synthesis of self-evidence and probability. Conviction contains self-evidence, because the knowing subject in the human sciences determines itself as a
spiritual subject, because knowing participates in the positing
of the known. It contains probability, because the spiritual
process transcends every individual positing and subjects the
arbitrary element in conviction to the judgment of history.
Both elements are one, however, in creative conviction, the act
of *spiritual self-positing*. Conviction has no relation to opinion
(though ordinary language frequently confuses the two words),
nor is it the same as self-evidence (as idealism sometimes seems
to say). Conviction accords perfectly with the creative nature of
spirit; it can no more be reduced to self-evidence and probability than creativity can be divided into the universal and the
particular.

Because of the influences of the human sciences on the other
areas of science, these other areas contain some degree of conviction. This is directly true in history, where exposition
depends on the comprehension of contexts of meaning; it is
also true in an indirect way, through the medium of the theory
of knowledge, in the other areas, where the pragmatic conviction of empirical experience is often supported by the creative
conviction of spiritual positing. Yet in these cases, there is the
same limit as there is in the application of systematics outside
the human sciences: conviction is based on the standpoint, not
on the relation between knowledge and the object. The fact
that conviction tends more toward self-evidence in philosophy
and more toward probability in the history of spirit does not
alter its nature: it never becomes either self-evidence or probability.

5. *The Human Sciences and the Spiritual Attitude*

Every creative act of spirit is, in its intention, directed
toward the *unconditioned form*. The unconditioned form does
not exist, however; it is the expression for the fundamental
relationship between thought and being. All thought is essentially directed to the comprehension of being. But being is infinite for thought; it is the abyss and the eternal Beyond of
every particular form. Thus thought creates an endless number

of finite forms, none of which exhaust being but all of which are subject to the unconditioned demand to grasp being. For being is the import, the reality, the unconditioned meaning that gives reality and meaning to every particular form. Therefore, every spiritual act of meaning-fulfillment bears in itself the Eros for the unconditioned meaning.

This will to the Unconditioned that is the foundation of spirit can find expression in two ways. It can seek to grasp being by means of the forms and their validity. We call this spiritual attitude *autonomy*, because in it the form is determined purely by itself. The first question here is not: How much existential import does a form express? The question is: Which is the correct form, the form that best satisfies the unconditioned claim of validity? The correctness and validity of a form reveal its power to grasp the Unconditioned.

The other possible attitude toward the Unconditioned is that spirit seeks to grasp the Unconditioned, or pure being, immediately. We call this attitude *theonomy*, because in this attitude the nature of all spiritual realizations is determined by the will to grasp the Unconditioned immediately through these realizations. Here the first question is not: Which is the most correct form? The question is: Which is that form that expresses the import of the Unconditioned most powerfully? We discuss the relation between the two spiritual attitudes, both with regard to logic and to the human sciences, in the foundation of the theonomous sciences. Here we establish the distinction because it is critical for the construction of the system of the human sciences.

This duality of spiritual attitudes presents the human sciences with the task of positing norms for both autonomous and theonomous meaning-giving. There are human sciences of both theonomous and autonomous spiritual realization. The structure of the system is determined by this distinction. Of course the two parts are not analogous. The antithesis between autonomy and theonomy is directely applicable only in those spiritual functions that are directed to the Unconditioned, metaphysics and ethics. There is a theonomous and an autonomous metaphysics and ethics. In the other areas of meaning, however, this disctinction is applicable only insofar as they are dependent on metaphysics and ethics. Considered in themselves, they are always autonomous, always directed toward valid forms. Thus there are no analogous theonomous functions corresponding to autonomous science, art, community, and law. Accordingly, we develop the system of the human sciences within the autonomous sphere, then examine the influence of the theonomous attitude on the individual areas of meaning.

The presupposition of this view is that *religion* is not one sphere of meaning alongside the others; it is an attitude within all spheres: the immediate directedness to the Unconditioned. When the unconditionality of the Holy has been grasped, there can be no question of classifying religion alongside the other areas or even of placing it above the others. Consequently, the normative science of religion cannot be concerned with one object alongside others; it is concerned with an intention that is possible within all the areas of meaning. The normative science of religion is the theonomous human science.

6. The Structure of the Human Sciences

So far, we have discovered two principles for organizing the human sciences: the elements of these sciences (philosophy, the history of spirit, and systematics) and the two spiritual attitudes (autonomy and theonomy). In the center of the structure, however, stand the organization of the spiritual areas and the human sciences corresponding to these areas. The elements and the spiritual attitudes receive their content only in the *objects of the human sciences*.

The organization of the spiritual areas is based on the functions of the spirit-bearing gestalts. For to realize spirit means to subject these functions to validity. This has led to the view that the organization of the areas of spirit must conform to the faculties of the soul, especially to the faculties of thought, volition, and feeling. This view is still found in Kant. But it cannot be carried out, for the organization of the *psychic functions* is quite uncertain, and the comprehension of the areas of meaning is, as an act of creative conviction, independent of such variations. Furthermore, it is difficult to make the number of areas of meaning correspond to the number of psychic functions — this is especially evident with regard to art and religion.

Ultimately, all psychic faculties participate in every spiritual act; the soul is uniformly directed to the contexts of meaning. Psychic functions are not spiritual functions, though they are the existential form of spiritual functions.

The individual gestalt becomes spirit-bearing when the universal, the thought form, is projected into this gestalt — that is, when the gestalt achieves freedom. Spiritual acts are those acts in which the individual gestalt establishes its relations to reality in freedom, or in a valid way. The dual act of every gestalt (i.e., the assimilation of reality and the insertion of itself into reality) is spiritual when it is a meaning-fulfilling act in both directions. The meaning-fulfilling act that assimilates reality is *theoretical*; the act that inserts itself into reality is *practical*. In the theoretical act, the spirit-bearing gestalt assimilates the forms of reality. The gestalt can contain things only as forms. Within the realm of being, however, things stand alongside things, gestalts alongside gestalts. In the practical act, the spirit-bearing gestalt establishes an existential relationship. But a free, meaning-fulfilling existential relationship is possible only between beings that have assimilated the universal — that is, between spirit-bearing gestalts. All other existential relationships lie in the prespiritual sphere of the gestalt life and can attain spiritual significance only in an indirect way, as we have seen in our discussion of technology. The establishment of the universal thought relationship and of the universal existential relationship: this is the fundamental dual act of the spirit-bearing gestalt. It is an act that is the basis for the division of spiritual areas into a theoretical and a practical series.

Within each of the two areas, there is another principle of division at work. The spiritual act can be directed to the individual form and to the individual being; it can also be directed toward the Unconditioned that supports all individuals and is the foundation of individual meaning-fulfillment. The functions directed toward the Unconditioned are the *supporting* functions, and those dependent on the supporting functions are the *supported* functions. Finally, the latter functions are subject to a further division: they can be directed either to the forms in which the theoretical and practical relationships are presented or to the import that these forms express. The first direction yields the functions determined by *form*, the second yields those determined by *import*. This division is applicable to both the theoretical and the practical spheres. Thus the thought relationship can be determined by import when the existential import of things is assimilated by the forms, and the existential relationship can be determined by form when the rational forms of such relationships are to be established.

These principles provide the following organization of the areas of meaning. In the theoretical sphere, the supported function determined by form is *science*, the supported function determined by import is *art*, and the function supporting them both is *metaphysics*. In the practical sphere, the supported function determined by form is *law*, the supported function determined by import is *community*, and the function supporting them both is *morality*. The significance of and justification for this organization of the spiritual functions can be demonstrated only by our examination of the individual human sciences.

Though the supporting functions of the theonomous spiritual attitude correspond to the supporting functions of the autonomous attitude (the *formation of myths and dogmas* corresponds to metaphysics, *piety* corresponds to morality), there are no independent supported functions in the theonomous sphere. Theonomy exists in them only to the extent that the supporting functions influence it.

Thus, the system of the human sciences is founded on three principles: the methodological, which yields the elements of the human sciences; the material, which yields the objects of these disciplines; and the metaphysical, which reveals the spiritual attitude. These three principles correspond to the three major alternatives of contemporary philosophy: the idealistic, or methodological; the realistic, or objective; and the metaphysical, which is concerned with the philosophy of life. They ultimately correspond to the three elements of science itself: thought, being, and spirit.

II. The System of the Human Sciences

A. *The Elements of the Human Sciences*

1. The Doctrine of the Principles of Meaning (Philosophy)

a. The Concept "Philosophy"

Introductions to philosophy usually begin with the statement that philosophy is the only science that assigns itself to itself as its own object. On closer examination, however, it is not philosophy but the *systematics within the sphere of science* that assigns their tasks to all sciences, including itself. Thus the statement about philosophy would be true only if the systematics in the sphere of science were itself a part of philosophy. According to our outline, this is not the case, for we distinguish between philosophy and systematics.

"Philosophy" happens to be the name not only of science but also of metaphysics. At some time or other it has included every part of science. Its range diminished when the particular sciences surrendered their connection with the whole, leaving only logic, psychology, metaphysics, and ethics. Kant has replaced metaphysics and ethics with the *critique of reason*; when psychology also became independent, like the other empirical sciences, only logic and the critique of reason remained within the domain of philosophy. This is basically still the contemporary situation, except that the systematic human sciences are partly included in philosophy and that there are new attempts to construct metaphysical systems, especially under the rubric "philosophy of history." Another factor that has changed the situation has been the appearance of the phenomenological method, which attempts to provide a philosophy that will be a *foundational science* for all the disciplines.

The following considerations are important for our definition of philosophy. The concept "philosophy" would lose all meaning and simply merge with the concept "science" if one repudiated the process by which the individual sciences became separated from philosophy, designating the *whole of science* "philosophy." This also applies to phenomenology, which we can regard, not as a foundational science, but as a method that continues to be useful in all the sciences. If philosophy is assigned a particular task, then the current division of labor within the faculties should not influence the formulation of concepts. Thus it is obvious that we must exclude psychology from the philosophical sciences and that we must establish the peculiarity of metaphysics as a special sphere of meaning standing alongside and above science. Furthermore, it is necessary to exclude logic from philosophy, for logic is related to the other sciences in the same way mathematics is: the two constitute the sciences of pure thought. Therefore, by "philosophy" in the strict systematic sense, we mean the doctrine of the principles of meaning, or the *doctrine of the spiritual functions and categories.*

One could object to this definition by saying that the doctrine of principles is so deeply rooted in logic, on the one hand, and in metaphysics, on the other, and that it develops so directly in the human sciences, that it is more correct to give this entire complex of labor the name "philosophy" and to subordinate the doctrine of principles to it. There can be no disagreement on this point as long as one is consistent and regards all the human sciences, including theology and the systematics of law, as philosophy. This terminology would agree with both tradition and linguistic sense even better than the limited use we are proposing. But the concept "philosophy" then becomes meaningless for scientific systematics, which has the task of analyzing complex sciences into their elements. For systematics it is necessary either to eliminate the concept "philosophy" completely or to restrict it to the doctrine of the principles of meaning. The broader use of language then allows logic, the human sciences, and metaphysics to combine under the name "philosophy."

Simmel introduces another use of the word "philosophy" when he defines philosophy as the investigation of things from the point of view of *totality*. But it is clear that both this definition and Simmel's use of it in his analysis represent a synthesis of logical and aesthetic elements in the form of metaphysics.

b. The Objects of Philosophy

Philosophy is the doctrine of the principles of meaning; that is, it is the doctrine of the meaning-giving functions and categories. Functions are those directions of the spiritual act by which the independent areas of meaning are delineated;

categories are the forms by which objects in the areas of meaning are constituted. The functions and categories of meaning are the principles on which all meaning-giving is based.

Functions of meaning are the directions of the act in which the spirit-bearing gestalt achieves its relation to reality in a valid way, thereby erecting a meaningful reality. Accordingly, the criterion for the independence and necessity of an area of meaning is the indispensability of the area for the construction of a meaningful reality. This is the deductive element that is inherent in critical understanding and that resembles the critical conviction of self-evidence. Without functions, the spiritual unity of consciousness would be destroyed and reality would become meaningless; functions have the certainty possessed by consciousness and meaning themselves. But this certainty is not simply self-evidence. It is the immediate certainty of life, the certainty a spirit-bearing gestalt has with regard to itself. The spiritual life is not fixed in any moment, however. The inner infinity of the spiritual process drives it beyond every one of its creative realizations. Even the unity of meaning in which a spiritual gestalt lives does not possess the unconditionality of pure form. The system of the functions of meaning is not an object that can be known in a self-evident way; it is an object known by conviction, however much conviction resembles self-evidence.

The function of meaning is neither the giving of meaning, as idealism contends, nor the grasping of meaning, as realism asserts. Spirit does not give laws to things, and things do not give laws to spirit. Idealism is wrong because it cannot show how the forms of meaning correspond to things; realism is wrong because it cannot show how things correspond to the forms of meaning. We have defined the relation between things and the forms of meaning as one of *meaning-fulfillment*. This means that things are directed to the unconditioned form and that this direction is fulfilled in spiritual creations. Neither ideal norms existing beyond being nor a meaning-formed reality confronting spirit are bearers of meaning. Meaning is not given, either really or ideally; it is intended, and it attains fulfillment in spirit. Every reality contains the intention to meaning-fulfillment, because every reality is directed to the unconditioned form. Of course spirit cannot know what the reality is in abstraction from meaning-fulfillment, for knowledge is one kind of meaning-fulfillment. But by negating all forms of meaning and metalogically descending beneath the sphere of meaning, spirit can apprehend that which precedes meaning-fulfillment: the "kingdom" of unfulfilled intentions, where not even the alternative existence/nonexistence is applicable. The full explanation of these ideas belongs within epistemology. Here we only want to show what we mean by saying that the functions of meaning are the objects of philosophy.

Categories are the forms of meaning that constitute the object in every area of meaning. Meaning, with its objects, is established by applying the categories to being in accordance with a function of meaning. Thus the categorial forms are presupposed in the comprehension of every object; but they themselves are not objects. Like the functions of meaning, these forms can therefore be grasped only through a critical understanding of the meaning-giving forms that are necessarily contained in every object. And like the functions of meaning, they are not self-evident. The forms of meaning-fulfillment are creative. They are grasped by a conviction that approaches self-evidence but never attains it.

We have already discussed the most important categories in science, because they are crucial for structuring the system of the sciences. The fundamental concepts are those expressing the essence of meaning itself: "thought" and "being." These concepts are the *elements of meaning*; they are the foundation for grasping the functions and categories of meaning. But they themselves are neither functions nor categories. They constitute neither an area nor an object of meaning; they constitute meaning itself. They are therefore applicable to all the areas of meaning, although they can be grasped in the particular areas only by the analysis of the meaning-giving function (in this case, knowledge). The duality, form of meaning and import of meaning, is fundamental for every function of meaning. This duality is not *a* principle of meaning; it is *the* principle of meaning itself. Logism wishes to make being one category alongside the others. It does not see that by doing this, it deprives thought of its import and destroys the nature of meaning. Alogism wishes to make thought one function

alongside the others. It renounces the unity of all the functions, thus losing meaning itself. Metalogism penetrates
beyond the functions of meaning to the elements of meaning
itself, finding in them the universal, tension-rich principle
upon which it can erect the system of the principles of meaning
and therefore the system of spirit as such.

Every area of meaning has its own categories. Thus, besides
the categories of knowledge we have developed here, there are
also the aesthetic, the metaphysical, the legal, the social, and
the ethical categories; besides the autonomous categories,
there are the theonomous ones. Each of these *kinds of
categories* establishes special objects. Because of this, alogism
has spoken of different realities existing alongside each other.
But either reality means only the object of meaning-giving — in
which case there are obviously as many realities as there are
functions of meaning. Or reality signifies the metalogically
basic element of all meaning, being — in which case all meaning giving is based upon the one reality, the existent that is
fulfilled in meaning.

There is therefore a philosophy of science, of art, of
metaphysics, of law, of community, and of morality; all of
them can be pursued in both an autonomous and a
theonomous way. There is no *philosophy of being*, however.
The philosophy of nature, of history, and of spirit are
disciplines within the philosophy of knowledge. They develop
categories of scientific meaning-giving and point to areas of
objects that are constituted by these categories. In the interest
of clarity, it is better to avoid these designations altogether,
especially since they are also used in a metaphysical sense. But
when they are used in this latter sense, we have a right to re-
quest that they refer specifically to metaphysics, not to
philosophy.

We must still avoid one confusion to which the logistic
method is particularly prone. *The doctrine of the categories of
spirit* is a part of the philosophy of science. This doctrine
establishes the categories of spirit, whatever the spiritual function. The aesthetic, legal, and social categories are therefore
not scientific categories of art, law, community; they are the
special categories of these areas, categories that are scientifically grasped. The philosophy of science is concerned only with

spirit as a reality in the scientific, theoretical sense; it is concerned only with the existence of spirit, not with its intentions toward meaning.

c. The Metalogical Method

The methodological attitude of philosophy is *critical understanding*. This general definition is insufficient, however. The present situation in philosophy necessitates a debate with the major methodological alternatives, so that the systematic classification of philosophy is also established from the perspective of method. The concept "critical understanding" excludes two antithetical possibilities: "criticism" without "understanding" and "understanding" without "criticism." The former is characteristic of pure critical philosophy, the latter, of pure phenomenology. In its discussion with pure criticism and pure phenomenology, contemporary philosophy must develop the method that corresponds to its nature as the doctrine of the principles of meaning; it must develop the metalogical method.[2]

Both our definition of philosophy and our explanations of the functions and categories indicate that the critical attitude is necessarily bound up with philosophy. If all meaning-fulfillment is directed to the unconditioned meaning, then it is impossible to comprehend the functions and categories of meaning unless every form of meaning is directed to the unconditioned form. The achievement of the pure *critical method* is that it makes the relations of the forms of meaning to the unity of the unconditioned form the principle of its deduction, thereby avoiding alogistic arbitrariness. Furthermore, through its clear distinction between the forms and the objects of meaning, this method prevents the forms of meaning from being raised to the status of metaphysical objects, and it prevents philosophy from becoming a science of transcendental hypostatizations. Finally, it shows that the relation between spirit and things is not one between an original and a copy, but is one in which spirit gives meaning to things. The critical method therefore has three strengths: as rationalism, it penetrates to the unity of pure form; as criticism, it seeks this form within the forms of meaning of the objects themselves; as idealism, it recognizes the meaning-giving significance of spiritual acts. In these three respects, the fate of philosophy as the doctrine of the principles of meaning is bound up with the critical method.

But it is clear that the critical method lacks that form of understanding characteristic of the human sciences. Observation of the *rational unity of form* prevents an understanding of the import of the individual forms, with regard to both the functions and categories. All functions are grasped from the perspective of knowledge; everything else is grasped within knowledge. And within the sphere of knowledge, the categories are understood only insofar as the pure rational form is present in them. Thus, all spirit becomes logicized and everything real becomes rationalized. The critical method does not understand the principles of meaning; it critically extracts their logical element.

In order to understand the principles of meaning, the critical method would have to grasp the elements of meaning, thought and being. Because of its logistic approach, however, this method can define being only as a category or as the limit of knowledge. It disregards the positive import of being, for it sees this import as a metaphysical hypostatization. But it is therefore unable to grasp the meaning of meaning; it cannot understand understanding itself. Thus this method regards meaning-giving as a subjective act that is concerned with the logical formation of a formless reality; it overlooks the intention of reality toward spiritual meaning. It loses the unity of subject and object, of validity and being—the unity given in the creative fulfillment of meaning. Pure criticism therefore cannot do justice to the claim of philosophy to be the doctrine of the principles of meaning. It lacks understanding and thus cannot grasp the essence of meaning.

Phenomenology is different. Its major concern is understanding. It wants to assimilate reality before offering any criticism. Phenomenology wants to apprehend the pure essences wherever it can find them. The intuition of prominent fulfillments of individual forms of meaning is an aid to this apprehension. An essence is grasped by the intuition of an exemplary essence—not only an individual essence, but also the essential relationships within which the individual essence stands. In this way, phenomenology attempts to establish areas of meaning and to construct the system of the principles of meaning. The phenomenological method is superior to the pure critical method by the fact that it gives a central place to understanding. It can do justice to the special character of the functions of meaning; it can grasp the categories in their peculiarity. But it cannot distinguish between the principle and the object of meaning, because it does not admit the comprehensive unity of the unconditioned form as a critical criterion. It does not have the critical approach for grasping the principles of meaning. Because the intention of all things toward the unconditioned form, and therefore the dynamic relationship between reality and spirit, is foreign to it, phenomenology must deny that spirit fulfills meaning; it must find meaning realized in the essences themselves. That which is ideally given is at the same time the normatively correct. This is why phenomenological philosophy is always normative systematics as well, and this is why it overlooks the inner tension between the doctrine of principles and the doctrine of norms. Because it lacks critical principles, phenomenology has an *alogical, dogmatic* character. The pure phenomenological method deprives philosophy of its independence from the normative disciplines, making the doctrine of norms heteronomously dependent on the individual creations of the history of spirit. With the loss of the critical element, philosophy is no longer the doctrine of the principles of meaning.

The *metalogical method* is based upon the critical method. But it transcends the latter by assimilating understanding into criticism. To understand a form of meaning is to grasp the elements of meaning itself, the elements inherent in the form. The first task of the metalogical method is thus to *elaborate the elements of meaning*. The accomplishment of this task provides an infinitely tension-rich principle that is crucial for understanding every individual form of meaning. To understand a form of meaning is to perceive the tension between the elements of meaning that are immanent in the form. In this way, one recognizes both the relation of all forms of meaning to the unconditioned form (as the critical method does) and the peculiar significance of every form of meaning (as phenomenology does). For the elements of meaning include the meaning-giving element and the meaning-receiving one: thought and being—or more generally, for all areas of meaning: *form and import*. The meaning-giving element drives toward the critical deduction from the unity of pure form; the meaning-receiving element produces understanding, or the intuition of the tension between the two elements. But the two are inseparable.

The metalogical method overcomes the *logism* of the pure critical method. The meaning-receiving element of meaning is intended not only by knowledge, but by all the functions of meaning. As soon as knowledge seeks to understand the tensions between the elements of meaning, it must likewise make all functions come to life in itself. Otherwise, being is merely the negative correlate of knowledge, and it is thus not understood. The metalogical method overcomes the *alogism* of phenomenology in the same way. Metalogical understanding is critical understanding. The dynamic tension between thought and being yields a critical principle that is not satisfied with any given form of meaning, but drives toward the absolute form, which is the goal of all creative meaning-fulfillment.

The metalogical method is, first of all, the method of philosophy. But because of the fundamental significance of philosophy for the other two elements of the human sciences, we can also say that the metalogical method is the *method of the human sciences* in general. For every construction in the history of spirit and systematics is based on the comprehension of the elements and principles of meaning. All the human sciences are founded upon their first element, philosophy.

2. The Doctrine of the Material of Meaning (The History of Spirit)

The principles of meaning are concretely realized in *history*. History is the arena of creative fulfillments of meaning; new creations can be born only from history. In every spiritual act, the direction toward the unconditioned form, the intention to the universal, proceeds through the conditioned forms, through the particular that has been realized in history. The creative possibilities of every individual spirit-bearing gestalt, whether psychic or social, is determined by its place in history. The task of the history of spirit is to understand the concrete norms of the historical process from the perspective of the principle of meaning in order to present the material to the doctrine of norms for its normative decision. The point of departure for the history of spirit is the principle of meaning; its goal is the norm of meaning. The human sciences proceed from the doctrine of principles through the doctrine of material to the doctrine of norms.

The history of spirit receives the material of meaning and understands it from the perspective of the principle of meaning; it uses receptive understanding. And it constructs the material with the norm in mind; it uses arranging construction. Its distinction from history proper follows from these statements. The history of spirit is interested neither in actual historical spirit-bearing gestalts nor in historical causality. It is neither biography nor the history of culture. The histories of science, art, metaphysics, law, community, morality, and religion belong within the *history of culture*. The latter becomes the history of spirit only when the realizations of meaning in these areas are understood without regard to the unity of the spirit-bearing gestalts and their historical contexts. Obviously, elements of gestalt description and of causal knowledge are included within the presentation given by the history of spirit. But these elements are not its goal of knowledge; they are only its means of presentation. In the history of culture, however, these elements are the goal of knowledge; the understanding of the types of meaning is its means of presentation. Frequently the history of spirit is forced to provide itself with the historical material that the history of culture was supposed to furnish. In those cases, it combines the descriptive and the constructive tasks. There can be no objection to this collaboration as long as one remembers the distinction between the two goals of knowledge.

In principle, the history of spirit is free from temporal and spatial connections. In its use of material, it is independent of historical contexts. The history of spirit can combine materials that are completely isolated from each other (either temporally or spatially) within the unity of a meaning type; it can separate material that is historically contiguous—even the spiritual elements of one single gestalt — when it arranges its material. It is determined only by the way it must understand and arrange a spiritual phenomenon from the perspective of the principle of meaning. Every function of meaning contains the tension between the basic elements of meaning. This tension produces the basic types of possible realizations of meaning, and it produces the normative demand for a balance between the elements of this tension. Thus we have the *coordinates* of all arrangement in the human sciences: on the one side, the function of meaning with the inner tension between its elements; on the other side, the norm of meaning as the ideal synthesis of the elements in tension; in between, the spiritual types as attempts to attain the norm — types that are sometimes determined by one element, sometimes by the other. A phenomenon is understood (in the sense of the human sciences) when it is arranged within these coordinates.

The history of spirit stands in the *service of systematics*. This relationship is not immediately apparent, however; it is both possible and necessary to grasp and to arrange the historical material in and for itself, according to its basic spiritual directions. The history of spirit thereby adapts itself to the process of cultural history. But it does not become the history of culture. The history of spirit cannot understand a phenomenon without proceeding from the principles of meaning or leading to the norm of meaning. When it is satisfied with merely establishing historical contexts, and when it abandons arranging construction, it is the history of culture, not the history of spirit. The latter always rests on systematics and serves it, directly or indirectly. Sometimes it resembles the history of culture, sometimes pure systematics, and it forms a transition between the two. But it is neither the one nor the other.

There is one point, however, where the history of spirit has a direct relation to time: in the bearer of the normative element himself. The norms from which he creates are historically given to him as the elements for constructing his own spiritual life. He can understand the history of spirit only because he himself is shaped by this history. Now, if the goal of arrangement in the history of spirit is the normative system, then the material of meaning receives the greater significance the nearer it stands to the systematizer, the more he himself is shaped by this material, and therefore the more powerfully it influences his own creation. This explains the peculiar *perspective* of the history of spirit, a perspective that is exactly opposite the spatial perspective and that allows all tendencies of the history of spirit that are grasped constructively to converge in the systematizer. Every construction of the history of spirit attempts to arrange these tendencies so that they achieve synthesis in the bearer of the construction, not only in principle, but also historically. Every system places itself at the intersection point of the developments of the history of spirit.

The perspective of the history of spirit is deceptive when it is not recognized as a perspective. Then it becomes a metaphysical interpretation of history in which one's own system is posited as absolute. The most prominent example of this mistake is Hegel's appraisal of his own system as the synthesis and completion of the entire historical process. This position confuses the metaphysical *interpretation of history* with the *arrangement of history* by the history of spirit. The former attempts to apprehend a pattern of events from the point of view of the ideal synthesis of the elements of meaning—a pattern that necessarily transcends one's own creative standpoint. The latter attempts, by means of the concrete synthesis of the individual system, to arrange the types of the history of spirit in such a way that one's own system appears as the solution. These two tasks are independent of each other, although there are mutual influences present in the fundamental attitudes of both. If they are confused, the result is the absolutization of an individual standpoint—that is, the attempt to bring the creative spiritual process to a standstill.

3. *The Doctrine of the Norms of Meaning (Systematics)*

a. The Doctrine of Norms and Philosophy

The human sciences are completed by their third element, the doctrine of the norms of meaning, or systematics. Filled with the material of meaning, the principle of meaning becomes the norm of meaning. In its intention, the principle of meaning is universal. It is valid for every phenomenon it supports within the area of meaning. The norm of meaning has passed through history; it is born in a definite historical place and thus has the concreteness and particularity of the individual spiritual creation. In its intention, the norm of meaning is not *universal*; it is *universally valid*. It seeks to be the correct, valid solution to the problems that the history of spirit presents on the basis of the tension between the elements of meaning. It seeks to be the concrete fulfillment of the abstract, universal principle of meaning. We thus distinguish between philosophy of science and the normative doctrine of science, between philosophy of art and the normative doctrine of art, philosophy of metaphysics and the normative doctrine of metaphysics, philosophy of law and the normative doctrine of law, philosophy of community and the normative doctrine of community, philosophy of ethos and the normative doctrine of ethos, and philosophy of religion (or the theonomous doctrine of the principles of meaning) and the normative doctrine of religion (or the theonomous doctrine of the norms of meaning). By calling the normative part "doctrine," we are indicating the concrete, practical claim to validity that is essential to the doctrine of norms. The norm seeks to give direction to activity in all the areas of meaning; as doctrine, it seeks to be transmitted to the student — and this pedagogical phrase clearly expresses its normative nature.

There is thus a tension between *philosophy and systematics*. Philosophy is directed to the universal, from the perspective of which every phenomenon is understood and every concrete norm is grasped in its conditionedness. It is the critical, variable element that is akin to thought and *ratio*. Systematics is directed toward the universally valid, from the perspective of which all other norms are denied and one's own system receives practical, normative significance. It is the positive, invariable element that is akin to being and the realization of life. Nevertheless, the two are not antithetical. Systematics is *also* critical, and philosophy is *also* positive. Systematics is based on philosophy, deriving from it the criterion for deciding among the various possible realizations of meaning. And philosophy can extract the principles of meaning only from a concrete, living meaning. Philosophy is just as dependent upon systematics as systematics is upon it. There is no abstract understanding of meaning without concrete realizations of meaning; there are no principles of meaning without a world of objects of meaning. It would therefore have been just as possible to begin our presentation with the doctrine of the norms of meaning and then to discuss the principles and material of meaning as the elements constituting this doctrine. It is not as though philosophy were rational, the history of spirit empirical, and the doctrine of norms creative. Rather, all three of them are creative — or better yet, the *one* science of spirit is creative, but in philosophy it is more critical, in the history of spirit more receptive, and in systematics more productive.

It is important to see the relation between philosophy and systematics in order to solve the *practical conflicts* resulting from the concretely normative nature of systematics. Every creative system expresses the actual relationships between a spirit-bearing gestalt (psychic and social) and existents. The construction of concrete norms is an act that forms life. It is creative only when it forms life. But if the construction of a norm is creative, its destruction implies a partial or complete destruction of life for the spirit-bearing gestalt. This explains why spiritual gestalts cling tenaciously to their creative conviction as to a function of their life; it explains the psychic and social conflicts between philosophy and the doctrine of norms, between individuals and communities — especially when individuals, as the bearers of "doctrine," must represent the conviction of the community. The conception of spirit as creative yields the following principles for resolving such conflicts and for applying the doctrine of norms to them. If the doctrine of norms abandons the critical element (philosophy), it becomes the mere presentation of available norms. It loses the direction to the universal and becomes just a *historical self-presentation*, a confession. If it still claims validity, it becomes heteronomous and deprives spirit of its creative birthright. If the doctrine of norms abandons the positive element (the construction of concrete norms), it becomes the mere presentation of rational forms; it loses the living relation to reality and becomes a *universal schema*. If it still claims to be a concrete norm, it becomes an abstract law and impedes the creativity of spirit. Both psychic and social gestalts are subject to either possibility. As far as the conflicts between the two possibilities are concerned, however, it is true that the prius of criticism is creation, that the destruction of one form of life is therefore spiritually justified only if it involves the creation of a new form, and that the resistance to criticism can either be creative or hinder creation. There is no a priori decision on this matter. The internal and external conflicts resulting from the tension between philosophy and the doctrine of norms are therefore unavoidable. They necessarily have a tragic character.

b. The Doctrine of Norms and the Empirical Sciences

In the doctrine of the norms of meaning, the goal of knowledge is the *system*. The foundation for this statement has already been given. Here we must defend the systematic character of the doctrine of norms against attacks from the sides of both the thought sciences and the empirical sciences — attacks affecting the very essence of the doctrine of norms. The one side understands a system to be the realized rationality of things; the other side denies the very possibility of a system, pointing to the infinity of the real. As an alternative, a third side demands the "open system." The doctrine of the creative character of the system is opposed to all three positions. The system is always *self-contained*, as the very concept "system" suggests. It need not be complete or perfect; in the human sciences, a perfect system is as impossible as an assertion wholly outside any system. In this sense, every system is open. But in its intention, every system is self-contained, or closed; every assertion, however aphoristic, belongs within a self-contained system. This in no way implies a demand to realize a rational system, however. Such a system would not only always be imperfect, but would also be impossible. For the meaning of things comes to spiritual realization in a creative way. The absolute system is the attempt to realize the unconditioned form as such. The *absolute system* is a scientific utopia; the individual system is scientific creation.

Finally, the desire to banish the system as such is caused by the empirical conception of the human sciences. This view overlooks the productive, convictional character of these sciences; it disregards the fact that the system represents a living relation to reality, a unity of meaning, a system. For in meaning-fulfilling creation, in a *living conviction*, there can be no assertions of meaning coexisting in an accidental way. Meaning stands alongside meaning within a context of meaning, and conviction alongside conviction within a context of conviction. An incompatible coexistence is really an antithesis, an annihilation of meaning, a destruction of conviction. Naturally, we are not speaking here in a psychological way, but according to principle.

Just as spirit is misunderstood as the absolute system from the side of the thought sciences, so it is misunderstood as the lack of system from the side of the empirical sciences. But spirit is living meaning-fulfillment and creative positing.

At various points in the discussion of the *empirical sciences*, we have established its common boundary with the human sciences. What is the relation between the two kinds of science from the perspective of the human sciences, especially its systematic element? The history of spirit provides the first answer to this question: the historical material provided by empirical science is appropriated by the history of spirit, it is understood from the perspective of the principle of meaning, and it is arranged with the normative concept in mind. The systematic synthesis arises, however, from this material. Thus, here there is a clear and important dependence of systematics on the empirical sciences. New historical understanding can be of decisive significance for productive understanding. But this dependence should not be overestimated. On the one hand, historical understanding itself is conditioned by creative conviction; on the other hand, history is influential primarily through the immediate formation it gives the spiritual gestalt. Consciousness of history is not necessarily knowledge of history. The norm is born from the tensions of the historically conscious, spirit-bearing gestalt.

Besides the question of the relation of the human sciences to history, there is the question of their relation to the *gestalt sciences*. The relation is a direct one in the cases of psychology and sociology, an indirect one in the case of biology. We have drawn firm boundary lines from the side of the empirical sciences, but the fundamental distinction does not exclude factual influences. Psychology and sociology seek to know the structural laws of the spirit-bearing gestalts, the laws in which spirit achieves realization. The structural forms are the bearers as well as the limits of possible meaning-fulfillment. From this one might conclude that every positing of a norm would have to consider the structural limits of the spirit-bearing gestalt and that there is thus a decisive dependence of at least a negative kind on the knowledge of gestalts. But this is not true; the method of the psychology of spirit is greatly mistaken when it thinks it can prescribe guiding principles with regard to content for creative meaning-fulfillment. Structural limits exist, of course; to destroy them would lead to meaninglessness. But it is impossible to prescribe laws to spirit on the basis of these limits, for spirit is essentially the transcendence of structural laws. Spirit does not destroy these laws; it posits something new in them, something that is not derived from them, either positively or negatively. Therefore, psychology and sociology do not directly influence the doctrine of norms, though they indirectly influence it. The influence of structural laws on the spiritual process can be observed in history. The task of the *psychology and sociology of spirit* is to separate the constant structural forms from the individual historical realizations. These two disciplines show how to distinguish between the constant and the variable moments of the historical process. This is very important for both historical understanding and the construction accomplished by the history of spirit. It prevents one from considering structural forms as norms and individual creations as constant structures. It enables one to isolate the material of meaning in a pure way. But it itself does not furnish the material of meaning; it clarifies but does not establish systematics. The methodological influence of empirical science on the human sciences is therefore mediated exclusively by the history of spirit.

B. The Objects of the Human Sciences

1. The Theoretical Series

a. Science

Every meaning-fulfilling act in which a spiritual gestalt assimilates reality is theoretical. The *theoretical* act is directed toward the forms of things, for only as "thought" can the individual gestalt assimilate the real without being destroyed, not as "being." But the direction of this act to the forms of things does not mean that it intends only the forms. It intends reality, both as form and as import. Every theoretical act strives for a real relation between spiritual gestalt and reality, a real relation of meaning-fulfillment. Thus this act can take two directions: it can be directed to the forms in order to grasp things within the forms as such, and it can be directed to the forms in order to grasp the import of things through these forms. It can be determined by either form or import. The act that is determined by form is knowledge, and its creation is science.

Science is necessarily directed toward the forms of things, as forms. Science is always *formal*, but it need not be formalistic. It need not forget that knowledge establishes a real relationship of meaning-fulfillment to reality. If it forgets this, it resembles the pure forms of logic, thus losing reality. But if it surrenders its formal character, it resembles the aesthetic comprehension of import, losing the distinctive forms of things. The metalogical method preserves the independence of science from both logic and aesthetics. It points to the element of import in the distinctive forms of objects. It can do this because it recognizes that science is a supported function of meaning. In the doctrine of the elements of meaning, the metalogical method demonstrates the dependence of science on metaphysics; for the comprehension of thought and being, form and import, is an act that transcends the contrast between form-determined and import-determined—it is a metaphysical act. The relation between *metaphysics and science* is not one in which metaphysical symbols decisively influence the knowledge of things. This heteronomy of metaphysics is the sign of a false, rational metaphysics. The relation is one in which the metaphysical grasp of the import of meaning permits an understanding of the principles of meaning; it is one in which the formation of metaphysical symbols depends on the scientific formulation of concepts. The doctrine of metaphysics will determine this mutual relationship between the supporting and supported functions more precisely.

Like all the human sciences, the science of science consists of the three elements of philosophy, the history of spirit, and systematics. The philosophy of science is usually called the *theory of knowledge*, or epistemology. Since Kant's critique of reason, epistemology has achieved a dominant position in philosophy. Replacing the rational metaphysics that had been destroyed, epistemology attained the rank of a fundamental science. The prejudice developed that all real knowledge depends on the theory of knowledge — a prejudice to which one could rightly object that even the theory of knowledge is already a knowledge for which there must be another theory, and so on. In fact, the situation here is the same as that in the other areas of meaning: the spiritual process necessarily occurs in consciousness, but not necessarily in scientific consciousness. This is also the case with the scientific process. This process does not rest on self-reflection, but on the certainty of its functional and categorial foundation. Even thought about the process itself depends on this certainty. Like every human science, however, this thought is productive; it influences the immediate process of knowledge.

Epistemology is the doctrine of the functions and categories of knowledge. It is the *philosophy of science*. For the sake of maintaining the analogy with the other areas of philosophy, it might be appropriate to replace the concept "epistemology," which is terminologically and materially questionable, with "philosophy of science." This discipline is not a theory of the origin of knowledge, however, and is thus not a problem belonging to the psychology of spirit or methodology. In fact, it is a doctrine of the principles for scientific meaning-giving.[3]

The systematic part of the doctrine of science, or according to the title of this book *the system of the sciences*, is concerned with the methodology and objects of knowledge. It deals with the material of meaning that the spiritual history of science presents to it; out of this material, it creates its normative idea of knowledge, both in general and in the particular areas. The critical criterion of this creation is the philosophy of knowledge, which is presupposed at every point in the system. The present book is essentially concerned with the structure of the system, considering objects and methods only insofar as they establish the general structure; but an exhaustive doctrine of science would treat all the methodological problems in the particular areas, however specialized the area. It would penetrate more profoundly into the actual problems of the particular sciences than was possible and necessary in our presentation. But it would never itself become a detailed system of all knowledge. A *detailed system* is an invasion by the human sciences into the life of science itself; it is comparable to an attempt to unite all artistic realizations within a system of art. Both attempts overlook the real existential relationship contained by every meaning-giving function, the relationship that continually drives the function to new fulfillments of meaning. The formal system of the sciences is related to actual knowledge in the same way that the system of aesthetics is related to actual aesthetic intuitions and gestalts; but the detailed system wants to take possession of knowledge and intuition themselves and to bring them to a conclusion through the systematic form. The formal system is creative; the detailed system is both rationaily absolute and empirically relative, and it is destroyed by this contradiction.

b. Art

The theoretical act that is determined by import is aesthetic intuition, and its creation is art. It is impossible to treat *aesthetics and art* separately. Of course, aesthetic intuition is more comprehensive than artistic creation; but aesthetic intuition receives its fundamental expression only in the work of art, and every aesthetic intuition of reality depends upon the creative fulfillment this intuition has found in the work of art. Though the separation may to some extent be justified, even from the perspective of the empirical sciences (i.e., for the psychology, sociology, and technology of both art and aesthetic intuition), it is inadmissible for the human sciences, because they treat the function of meaning as such.

Science is directed toward the forms of things, as forms; aesthetic intuition attempts to grasp the import of things *through* the forms. Science is always formal, without necessarily being formalistic; art is always concerned with import, though it is unable to destroy form. The spirit-bearing gestalt can grasp the import of the real only through the form. A formless comprehension of import would also destroy the distinctive form of the bearer of aesthetic intuition. This would occasion the transition to the pure import that absolute *mysticism* seeks. But aesthetic intuition wants to grasp individual things directly, and it wants to grasp them through their forms.

Art and science proceed from the same material of reality. This is the reality that confronts the fulfillment of meaning but is directed toward this fulfillment. Thus we have the peculiar relationship between artistic and scientific forms: on the one hand, the material is identical, on the other hand, there is an absolute difference between the principles of meaning through which objects in both areas are constituted. And thus we have the continual violation of the boundary from both sides: the logicizing tendencies of art, especially in its realistic movements, and the aestheticizing tendencies of science, especially in the romantic view. In our own position, the boundary is clear: science seeks to grasp things from the perspective of thought, of pure form, without losing being, or import; art seeks to grasp things from the perspective of being, of pure import, without relinquishing thought, or form. In the first case, the form is subject to the law of form; in the second, it is subject to the law of import. And in the first case, what is important is its truth of validity; in the second, its truth of expression. The truth of science is correctness; the truth of art is power of expression.

This distinction also settles the doubts that could be raised against classifying the aesthetic sphere within the theoretical series of the functions of meaning. For us, *theory* does not of course mean knowledge, it means the meaning-fulfilling absorption of reality; it means ϑεωϱια, in the sense of the pure intuition of objects. But in this sense, science and art belong together in the same series.

This classification is contested from two sides. The aesthetic is placed over against the theoretical and the practical as a special function of meaning: then it is either associated with feeling as a psychic function or it is regarded as the synthesis of the theoretical and the practical. The first view is subject, to begin with, to the fundamental objection we have raised to the association of the functions of meaning with the psychic functions in general. Every meaning-fulfilling act contains *feeling*; a definition of the aesthetic from the perspective of feeling would have to emphasize the distinction between aesthetic feeling and the other feelings. But that cannot be achieved by referring again to feeling. Feeling certainly plays a special role in the aesthetic attitude. For in feeling, the individual gestalt becomes conscious of the existential relation to things. Precisely for this reason, however, one cannot define the aesthetic from the perspective of feeling, perhaps saying that art creates "manifest symbols of feeling." This definition contains a subjectivism of feeling that knows nothing of the individual import of things and does not see that things can become symbols only because feeling grasps their individual import in a spiritual, meaning-fulfilling way. A formalism and rationalism in the conception of things correspond to the subjectivism implicit in the definition of the aesthetic as feeling. But when the metalogical method sees the import of objects in every form of things, the subjectivism that views the aesthetic as feeling loses its justification.

It is just as wrong to see the aesthetic as a *synthesis of the theoretical and the practical* and to regard art as the representation of the ethical ideal in visible form. This view depends on the doctrine of the primacy of practical *reason*, which, like the definition of art as feeling, rationalistically overlooks the individual import of things. Similarly, this view is determined by a classical aesthetics demanding that art be the realization of ideals. Art ought not to realize ideals, however; it should make essences visible — even essences that are morally harmful, from an ethical viewpoint. The criterion of art is its power of expression, not the ideal nature of what is expressed.

The flaw in both views of the aesthetic is based on the fact that they forget that metaphysics supports art. They are thus unable to recognize an *existential import of reality* that is just as independent of subjective feeling as it is of rational form; they are forced to attempt to classify the aesthetic alongside the theoretical and the practical, though they cannot do so.

Art is therefore a supported theoretical function. The existential import of things that it attempts to grasp is the revelation of pure being, of the unconditioned import within the particular forms of things. The entire direction of aesthetic intuition must be shaped according to how being is apprehended and symbolized in the basic metaphysical attitude. The influence of the metaphysical attitude on aesthetic form is *style*. Whether it be the style of a period, a group, an individual, or even a particular work of art, style is the universal determination of aesthetic forms by the way import in general is grasped. Just as the dependence of science on metaphysics is evident in the grasping of the elements of meaning, thought and being, so the dependence of art on metaphysics is evident in the tension between form and import, that is, in style. Art without metaphysics is also without style. Such art is either abstract formalism or formal arbitrariness.

Aesthetics, the human science that is concerned with art, is composed of the three elements of the philosophy of art, the spiritual history of art, and the doctrine of the norms of art. Philosophy of art examines the aesthetic functions and categories. In particular, it must articulate the difference between the aesthetic and the scientific categories, though it cannot maintain that aesthetic objects have a special ontological status. The spiritual history of art is essentially the spiritual history of style. It expounds the different tensions found in art between form and import; it leads to the ideal synthesis, to the balance of the tensions, which is treated by normative aesthetics.

c. Metaphysics

i. The Present Situation in Metaphysics

Kant's *critique of rational metaphysics* still dominates the present situation in metaphysics. Today hardly anyone is demanding a revival of this obsolete form of metaphysics. The critical attitude toward all metaphysical endeavors is still in our blood. It is directed to both pre-Kantian and post-Kantian metaphysics. The critical attitude interprets both forms as attempts to know, by scientific means, objects that transcend science. Even the strictest critical attitude cannot avoid some metaphysical elements, of course; every meaning-fulfilling act contains some intuition of the elements of meaning, and the comprehension of these elements is always metaphysics.

The critical attitude is justifiably directed against the attempt to place metaphysics on the same level as science. When the Enlightenment made metaphysics a rational science, it completely abandoned the metaphysical attitude. It attempted to draw the Unconditioned down into the sphere of the conditioned, into the sphere of proof and disproof. But in this way, metaphysics was deprived of its object even before its work began. Kant's critique merely drew the conclusion from this state of affairs.

The critical protest against regarding *metaphysics as a science* must be directed not only against the rational kind of metaphysics, but also against the empirical kind presently demanded. Metaphysics is considered a dubious, though desirable, conclusion of the individual sciences; it is viewed as the summary of the most universal and therefore least certain hypotheses of the other sciences in one picture of the world. It functions on the same level as the individual sciences, but it is uncertain, because it oversteps the limits of methodological experience. A rational probabilism replaces rational dogmatism. But there is no reason the most universal hypotheses should be called "metaphysics" while the less universal are regarded as science. Neither of the two is metaphysics.

The critical point of view is thus correct to the extent that it opposes the classification of metaphysics within science. But it is incorrect when it then rejects metaphysics as such, for there is no spiritual attitude without metaphysics. The present situation in philosophy confirms this judgment from the most diverse sides. On the one hand, critical philosophy itself, in its doctrine of the *antinomies*, constantly encounters the problem of the Unconditioned; it increasingly recognizes the necessity of confronting this problem directly. On the other hand, the phenomenological school, in pursuing its logical realism, is to some extent moving toward a new *ontology*. At the same time, the individual areas of science are moving toward a more profound grasp of the elements of meaning. Whenever the *existential element* (as opposed to the rational form) is present in things, there are trends toward metaphysics. Finally, the entire historical situation of Western culture, especially the fundamental critique of culture, has emphasized the problems of the *metaphysics of history*. Though ontology still lacks a certain confidence, hardly anyone questions the necessity and the justification for a metaphysics of history. Thus all aspects of the spiritual situation tend toward a new positive attitude to the problem of metaphysics.

Here we can say nothing about the tendencies in *art* that are moving toward a new metaphysical attitude. These tendencies are just as strong as those in science, however, and they are even clearer and more conscious.

ii. Metaphysics as an Independent Function of Meaning

The task of the doctrine of the principles of metaphysics is to show that the direction of consciousness toward the Unconditioned is a necessary function that constitutes the reality of meaning. It must show that the prius of every individual comprehension of meaning is the unconditioned meaning itself, that the prius of every form of meaning is direction toward the unconditioned form, and that the prius of every import of meaning is the unconditioned import. The concept *"the Unconditioned"* is therefore the central metaphysical concept. Metaphysics is the will to grasp the Unconditioned. It does not ask whether the Unconditioned exists. This question is meaningless, because it already presupposes the Unconditioned of meaning as the context of meaning. If the Unconditioned were established, it would no longer be the Unconditioned; it would be some object whose existence it is possible to prove. The Unconditioned cannot be proved; it can only be indicated as the meaning that supports all meaning-fulfillments.

The direction toward the Unconditioned is neither scientific nor aesthetic intuition. Reality is neither to be grasped in the individual forms of objects, as in art. Rather, the Unconditioned itself is to be grasped — the Unconditioned supporting every ed itself is to be grasped — the Unconditioned supporting every individual form and every individual import. Metaphysics is concerned with a self-sufficient giving of meaning that is basically independent of scientific and aesthetic comprehensions of the world. Yet metaphysics has no other forms of expression at its disposal than those presented to it by scientific and aesthetic intuition. The formal means of expression of metaphysics is the concept. This is the *scientific element* in metaphysics; the constant transgressions of the boundary between science and metaphysics are based on this fact. But metaphysics uses scientific concepts in their aesthetic sense. The intention is not directed toward concepts insofar as they are forms of reality, but insofar as they express the import of reality. The *artistic element* inherent in every creative metaphysics is based on this fact. Thus metaphysics contains both aspects, the scientific as well as the artistic, with equal justification and with equal significance. It is usually a sign of decline when one of the two elements displaces the other. But the metaphysical attitude is not a synthesis of the two others; it supports them. This attitude is the original unity from which the other two have proceeded and to which they essentially strive to return. One can still see this clearly in theonomous metaphysics at its mythological stage.

Metaphysics attempts to grasp the Unconditioned. But it can grasp the Unconditioned only in the forms of the conditioned. This is the profound paradox inherent in metaphysics. From this paradox, we see the only way scientific concepts can be used by metaphysics. If one calls a concept that expresses something other than its proper, immediate meaning a "*symbol*," then all metaphysical concepts must be designated symbols. Metaphysical concepts are expressive and therefore valid, though they are not valid in the way scientific concepts are. What they express is not a subjective feeling, nor is it the individual import of reality, as it is in art; they express absolute import.

The central symbol for the Unconditioned is the unity of the real itself. The way this unity is seen, the categories from which it is constructed, the valuations expressed in its construction — all of these yield the *system of metaphysical symbols*. But the symbolic character of metaphysics does not imply arbitrariness. Rather, the selection of symbols is based upon two factors: on the one hand, the choice is determined by the metaphysical attitude, and on the other, it is determined by the conceptual material that science, especially philosophy, presents for the formation of symbols. Certainly metaphysical symbols claim to *adequately express* that which is to be represented by them as the essence of the Unconditioned, and the metaphysical attitude itself has the degree of certainty corresponding to the unconditionality of its object. Nevertheless, metaphysics cannot create self-sufficient symbols. It depends on the conceptual formulations of science. If, as metaphysics, it attempts to form concepts claiming scientific validity, it loses its character as directedness toward the Unconditioned. It becomes science and heteronomously influences science with the claim to unconditionality that is essential to it. Metaphysics is crucial for both science and art in only one respect. It supports the comprehension of the elements of meaning in both of them. It underlies the method of knowledge and the artistic style. Whether or not it is recognized as a function, metaphysics provides both the import and the fundamental intention to both scientific and artistic meaning-fulfillment.

iii. The Structure and Method of Metaphysics

Our task here is not to present a complete normative *doctrine of metaphysics*. But since both metaphysics and its justification as an independent function of spirit are contested, we must say a few things about the nature and procedure of metaphysics.

The first and fundamental task of metaphysics is the *doctrine of the elements of meaning*. When its independence from science is established as a fact, metaphysics cannot attempt to grasp the Unconditioned from the perspective of being, but must try to grasp it from the perspective of meaning. If the Unconditioned were an existent, an object alongside other objects, it could be investigated in a scientific way. But the Unconditioned is the meaning that supports all meaning-giving, the meaning that can never be made into an object. Thus the only way to grasp it is by apprehending the elements that constitute meaning.

The system of metaphysical symbols is based upon the foundation of the elements of meaning. Every metaphysics must answer three basic questions: the question of the relation of the Unconditioned to existents; the question of the relation of the Unconditioned to the creative spiritual process; and the question of the meaningful unity between the existential and the spiritual processes. These three questions are not questions concerning being, but questions concerning meaning: genuine metaphysics is the *metaphysics of meaning*.

The metaphysics of being, or ontology, answers the first

question. Ontology must show how being as a whole, as a universal gestalt, is a symbol for the unconditioned meaning. It must focus, constructively and critically, on those aspects of reality in which the unconditioned meaning, the ultimate unity of the elements of meaning, is most completely expressed. The task of ontology is therefore not that of apprehending an existent behind phenomena. Its task is to represent the structure of all existents and their unity as an expression of pure meaning.

The distinction between the metaphysics of being and the *metaphysics of history* corresponds to the contrast between an idea of a universal gestalt and an idea of universal history, a contrast we encountered in our discussion of the empirical sciences. The metaphysics of history is the interpretation of the meaning of the spiritual process from the perspective of the unconditioned meaning. Its task it not to apprehend a mystical spiritual substance standing behind history; its task is to see the revelation of the unconditioned meaning in an interpretation of the meaning of the historical process. But both the metaphysics of being and the metaphysics of history unite in the investigation of the universal process, in which the contrast between the intention of meaning and meaning-fulfillment is also conquered. Only the union of the two creates the ultimate, highest symbol of the Unconditioned, the ideal unity of the elements of meaning, a unity that is both the goal and the ground of all being and becoming. The system of metaphysics culminates in this *metaphysics of the absolute idea*.

Metaphysics can be called the doctrine of world views. But if we use this term, we must divest it of everything suggesting subjective opinion or scientific hypothesis. If "world view" is to be the concept that combines metaphysics and ethics, or the theoretical and the practical attitudes toward the Unconditioned, then it is the fundamental act of spirit as such. It is the foundation of all theoretical and practical functions, and it contains the unconditioned seriousness and the unconditioned responsibility that are appropriate to directedness toward the Unconditioned. In view of the subjective connotation the phrase "world view" has acquired in ordinary language, it is perhaps more correct to call the unity of metaphysics and ethics in the psychic gestalt a *"spiritual attitude"* and in the social gestalt a *"spiritual situation."*

Because the symbol is the means of expression in metaphysics, one can also speak in a heterogenous sense of a metaphysical method. The goal of knowledge in metaphysics is the unity of *being* and *meaning*, or the system, which is both a universal gestalt and a universal framework of meaning. The cognitive attitude in metaphysics is the unity between the comprehension of form and the comprehension of import, between scientific and aesthetic intuition, between the *perception of being* and the *understanding of meaning*. The procedure of knowledge in metaphysics is the intuition of the unconditioned import within the conditioned forms; this method was called the "contemplation of the *coincidentia oppositorum*" in Renaissance philosophy, and it is essential to the metaphysical procedure. Concepts such as "intellectual contemplation," "pure intuition," comprehension of "the absolute identity" and of "the paradox" express the *method of coincidence*. But we must emphasize that these concepts describe the metaphysical method, not a scientific method; they are fatally misunderstood when the independence of the metaphysical function and its forms of expression is not recognized.

The kind of certainty found in metaphysics is a unity of unconditioned certainty and conviction. The direction to the Unconditioned contained in every metaphysical statement has the unconditioned certainty of the Unconditioned, a certainty that supports all self-evidence, even if it is formal. On the other hand, the forms of expression in which this direction is represented have only the nature of conviction. They are spiritual creations; they possess the kind of certainty contained in every spiritual creation, scientific or artistic. This peculiar, tension-rich union of original certainty and conviction is essential to every metaphysics.

2. The Practical Series

a. Law

i. Law as a Function of Meaning

In the theoretical functions, the spirit-bearing gestalt assimilates the real into a meaningful thought relation, but in the practical functions, it establishes existential relationships to the other spirit-bearing gestalts and thereby to reality in general. Practical meaning-giving is the *realization of meaningful existential relationships*. Psychic or social gestalts in all spheres of meaning can be spirit-bearing. Yet there is a difference between theoretical and practical functions. It is possible to imagine theoretical functions supported only by psychic gestalts. Of course, this notion abstracts from the fact that every psychic gestalt is a member of a social gestalt; it overlooks the actual dependence of even the theoretical functions on the life of the social organism. In the practical functions, on the other hand, this abstraction is impossible. A meaningful existential relationship is always supported by *social relationships*. Every personality (i.e., every individual psychic gestalt) who is a bearer of spiritual existential relationships exists within the community; the practical consciousness of validity appears only when he encounters the resistance offered by the other individual members of the social organism. Without the existence of other spirit-bearing gestalts, the existential relationship would be subspiritual, remaining purely in the gestalt sphere. Therefore, meaning-fulfilling existential relationships of the individual to things and to reality are always socially conditioned relationships within which the socially formed and socially determined personality acts. There is a spiritual existential relationship only for the socially integrated psychic gestalt, never for the socially isolated one, regardless of the fact that the latter gestalt is an abstraction.

Just as in the theoretical sphere things can be grasped either in their forms or in their import by means of the forms, so in the practical sphere existential relationships can appear either in *combinations of form* or in the *community of existential import*. This contrast is not absolute, however. Combinations of form are supported by a common import, and the community of import must appear in meaning-fulfilling forms. The difference is only that in the first case we are concerned with the forms as such, but in the second case we are concerned with the forms only insofar as they express the community of existential import.

Law is the form-determined existential relationship within the spirit-bearing social gestalt. Both form-determined functions are subject to thought; both stand at the rational pole of the spiritual functions. Rational enforceability, in law, corresponds to rational self-evidence, in science. Scientific form exercises the same control over all subjective arbitrariness in knowledge that legal order exercises over all subjective arbitrariness in action. Therefore enforceability, or *rational compulsion*, is essential to every legal proposition, but both the aesthetic and the immediately communal attitudes are exempt from rational control. But the more science attempts to grasp being within things, the less it is determined by pure rational form. The rational compulsion of the legal form is limited to the degree that law attempts to be a living expression of communal import. Consequently, the formal nature of law must not lead to *legal formalism*. Legal formalism is as misguided as is scientific formalism; it attempts to realize the pure form of thought by destroying the living, existential reality. It isolates the individual with the principle of formal justice, or equality; then it attempts to establish unity, instead of proceeding from the living gestalt and adapting the legal form to the communal function of the individual.

Opposite legal logism stands *legal alogism*, which can manifest itself in various ways. Alogism can deny the normative character of law as such; it can explain law as the expression of the actually existing social relations without admitting any critical criterion. It can contend that law possesses real validity only by the power through which it is enforced. In connection with sociological technology, alogism can also conceive of law as the appropriate means for advancing the social organism, thereby giving law a validity relative to technology instead of the absolute validity we find in the human sciences. Both cases abolish law and an independent function of meaning, leaving the decision about its existential relationship either to the structural law of the gestalt sphere or to spiritless utility. The existential relationships of spirit-bearing gestalts lose their meaning-fulfilling nature, the law of meaning disappears, and social gestalts are dominated by the law of being.

The metalogical conception of law attempts to realize existential import within legal forms. It recognizes that law is a supported function, that it *depends on ethos*. Metalogic transcends both legal logism, which destroys community, and

legal alogism, which destroys spirit. It gives the entire legal
system an import derived from the will to realize the Unconditioned within existential relationships. A law that is not founded on ethos oscillates between formalism and arbitrariness.
The "correct law" is the import-filled law that is founded in
ethos.

Law is therefore the form-determined, supported function
of meaning within the practical series of the spiritual functions. Law is a necessary principle of meaning; it conditions the
reality of meaning. Just as one cannot reduce logic to
aesthetics, so one cannot reduce law to society. The legal form
(i.e., the rational, existential relation that is enforceable by
coercion) would remain, even if actual use of force were un-
necessary. The rational enforceability of law — its compulsory
nature — is not an empirical feature of law that could be overcome; it is a feature that is essential and always valid. The
distinctive form of the personality is maintained within existential relationships only by means of law; but in a basically
lawless community, all forms of personality would be in-
distinguishable, just as all forms of things would be in-
distinguishable in a reality that was only aesthetically, not
logically, grasped. Law is a *constitutive function of meaning*
and a necessary element in every meaningful reality.

ii. The Doctrine of Law

The normative science of law is composed of the philosophy
of law, the spiritual history of law, and the doctrine of legal
norms, or legal systematics. The latter requires special consideration. For though in the theoretical series of spiritual
functions, the creative character of the norm is ineffective as
long as there are no conflicts between social and personal conviction, in the sphere of law, the tension between both
elements of creativity is always present, because the legal
system has practical validity and demands unconditional subjection. The legal system is in operation because of the power
that posits law, whether or not this system corresponds to the
legal convictions of those subject to it. It has the sacredness of
validity for them, even when it contradicts their idea of justice.
This situation can yield two conceptions of legal systematics,

both of which deprive law of its creative character. The position of *natural law* begins with an idea of universal and rational law; it attempts to criticize actual laws from the perspective of this idea and to subject them to the pure norm of natural law. Natural law is critical and revolutionary. But it contains this dialectic: either it approximates the pure rational form (in the manner of legal logism), in which case it loses all concrete contents; or it contains a series of concrete legal contents derived from creative legal systems, in which case it can no longer claim to be universally valid. As soon as natural law takes the form of a concrete system, it becomes creative.

On the other hand, there is the position of *legal positivism*. In its romantic form, it maintains that every concrete law is the expression of the soul of a people and that there can therefore be no comprehensive idea of law. The task of legal systematics would not be that of criticizing from the point of view of a normative idea, but of systematizing existing legal intentions in a formal way. Legal positivism overlooks the fact that the soul of a people is a structural reality that is prespiritual and that can be spirit-bearing only by directing itself toward the universal. The creation of law implies the creation of correct law. The spiritual nature of law is destroyed when the creation of law is directed to the actual rather than to the valid.

But this discussion of fundamentals does not solve the problem of the doctrine of legal norms. In the theoretical sphere, the concrete norms from which systematics creates can be taken from the entire history of spirit; but the doctrine of legal norms depends on *current law*, in the further development of which it is engaged. Its task is not to construct a normative legal system from the laws of all peoples. Instead, the doctrine of legal norms must limit itself to demonstrating its normative intention toward a concrete law. This limitation is based upon the way the law is realized. The legal norm is born from the life of a concrete community; it is based on the will of the community that possesses legal power, the state. Effective legal conviction is therefore never the theoretical conviction of an individual; it is always the practical conviction of a community. The legal will of the majority or of a powerful minority can be expressed in this *practical conviction*. Seldom can this legal will unite the individual legal convictions of everyone. Every positing of law is a political act, that is, an act in which the tensions between powers contribute to the formation of legal realization. Thus the doctrine of legal norms can be creative only when it has a voice in the political will, when it influences legal practice, custom, and positing. But that is possible only when it makes its demands from within the concrete legal situation.

From this follows the first task of legal systematics: the normative interpretation of law. In and for itself, the understanding of given laws is a task of philological interpretation. But insofar as these laws are current, the interpretation has a normative character. This produces the distinctive phenomenon of *normative exegesis*, a discipline that is both historical and systematic. The historical nature of normative exegesis is expressed by the fact that the will of the legislator must be recognized by empathetic understanding; the normative nature is effective by the fact that through systematic construction, the spirit of the law is adapted to the spirit of the emerging legal conviction. If the tension between the two has become so great that the normative interpretation publicly contradicts the historical one, the time has come for positing new laws on whose formation the doctrine of legal norms can have decisive influence.

This distinctive relation between legal systematics and the given law cannot be overcome, even by revolutionary acts. A *revolution* can bring a new legal conviction to power, and the bearers of this conviction can transform the current law. But even revolution cannot produce a rational new creation in the sense of natural law, because a revolution cannot posit anew the actual life of the social organism.

The creative character of legal systematics would not be lost even if the formation of law were to transcend national boundaries and unite humanity in one law-bearing community. Even the creation of universal law would remain individual creation. *Universal law* would be a further development and transformation of existing laws, but it would not be natural law; even in this universal law, the task of the doctrine of legal norms would consist in the development of the current law toward new creative realizations of the idea of universal law. Thus, legal systematics provides impressive factual evidence for the creative nature of the normative human sciences.

iii. Law and the State

Law is realized by the *legally powerful community*, the state. Every act of positing law is based on a creative decision. In the theoretical sphere, the decision is reached by the free self-determination of spirit-bearing individuals; during the course of this self-determination, the motives that conflict with the conviction are either included or eliminated. In the practical sphere, the decision is made by the free formation of a consensus within the social gestalt; during the course of this formation, the individual wills that are in conflict with this consensus are forced into subjection. The rise of the will that posits law depends on both the legal intention and the tensions between the powers within the social organism. All legal realization is therefore based upon a creative decision of the law-bearing community.

The conception of the state as a community with the power to enforce laws contains the necessary interrelation between the state and law, but in such a way that neither can be reduced to the other. A power relationship becomes a state only by appearing in legal forms. Even when rulers place the law in the service of government, there is a recognition that law is a valid function—and this is what is important, not the subjective intention of the rulers. The state does not create law; the state becomes the state through positing law. The state is the *bearer of the decision* behind every realization of law, but it is not the bearer of the law; for law is an independent meaning-fulfilling function.

But the state cannot be reduced to law, contrary to the contention of the liberal theory of the state. No abstract legal organism could exist independently of the state. The prius of every realization of law is the community with the power to enforce laws, however this power is constitutionally divided. Both the creative character of law and the necessity for constant legal decisions make it impossible to realize law without a community with the power to enforce laws. The state is not merely a *legal order*; it is a community with the function of realizing law. Therefore, there can be no *law outside the state*. Every positing of law in a territory that had not previously contained a state involves the formation of a state, whether the law-bearing community is, in addition to being political, a community with a common goal, a religious community, a racial community, or a national community. Every one of these communities can creatively realize law and thus be a state.

This means that the doctrine of the state is part of the doctrine of community, not of the doctrine of law. Like every community, however, the state can be the object of legal formation and thus the object of the doctrine of law. The *doctrine* of constitutional law is the application of the idea of law to the law-bearing community itself. But the realization of constitutional law also depends on the creative decision of the political community. No constitutional law could be exempt from this rule, even if the law-bearing community were mankind itself. Even then, the tensions of power would be both the foundation of the legal decision and the creative substance of the realization of law. The state is the prius of constitutional law, not conversely.

b. Community

i. Community, Sociology, and Law

Community is a supported existential relationship that is determined by import. It is the form of meaning containing the common import of spirit-bearing gestalts that are related to being. Community is a form of meaning: this distinguishes it from sociological forms. And it is a form of meaning that is determined by import: this distinguishes it from law.

The task of sociology is to grasp the *existential forms* of social gestalts; the doctrine of community is concerned with the *forms of meaning* of communal relationships. Sociology must limit itself to the existential forms, describing and arranging them. In accordance with the structural method, it can erect forms that are ideal types. But it may not make these forms ideal and valid. Sociology is bound to be the existential forms and may not claim to determine the form of meaning by grasping the existential form. The sociological form of "the family," for example, is completely different from the social form of meaning of the familial community.

The doctrine of community has to discover the categories of the communal function. It has to establish, in the manner of the history of spirit, the directions in which the elements of meaning of the communal function have been realized; it has to indicate, in a normative way, how they ought to be realized. So the doctrine of community transcends every sociological structure, even though it must acknowledge that communal relationships can never be realized outside the confines of the sociological structures. In accordance with the universal law of spiritual realization, it transcends the sociological form; but it does not destroy this form.

Community is a function of meaning that is determined by import. Its forms therefore do not give existential relationships a rationally compulsory import; instead, they express an existential import. Their meaning is fulfilled, not by means of rational correctness, but by their power to express communal relations. Thus *communal forms* are distinguished from *legal forms* as clearly as aesthetic forms are distinguished from scientific ones. As one might expect, philosophical logism is unaware that the entire sphere of community is an independent area of meaning. For logism, there is only the formation according to law, on the one hand, and formation of the inner life of individuals, on the other. For it, the import-filled forms of communal life that are outside law are not independent meanings; they are merely the material for legal formation. This corresponds to the dissolution of being into thought, of the gestalt into atomic individuals and thus to a doctrine of categories oriented to mathematical natural science. In the theoretical domain, it would be like regarding aesthetically intuited reality as material for the scientific formulation of concepts.

The legal form that is rationally determined is *justice*; the communal form that is determined by import is *love*. Love is realized in forms that are just as inaccessible to rational enforcement as aesthetic intuitions are to scientific formation. Logism does not recognize the independence of the communal sphere, because it does not recognize being (as distinguished from thought) and love (as distinguished from justice).

Yet it would be wrong to reduce law to community, as alogism does. The rational character of law guarantees the independence of individuals who stand within existential relationships. It prevents both the *disintegration of individual gestalts* into formlessness and the transformation of love into injustice. Just as aesthetic intuition can never be separated from the objective reality grasped by knowledge without completely losing the gestalt, so the communal forms can never be separated from the rational law that protects the distinctive form of personality.

ii. Community and Personality

The doctrine of community is the doctrine of the norms of communal relationships. But these relationships are necessarily borne by personalities; the spirit-bearing psychic gestalt is constitutive for communal relationships. The doctrine of community is thus always the doctrine of personality as well. Community is the *community of personalities*; this is implicit in the concept "community." The concept "doctrine of community" necessarily includes the doctrine of personality. But since the prius of all meaningful existential relationships is social association, we speak of the doctrine of community, using this notion to include both community and personality.

This use of the term "doctrine of community" deviates from the usual conception, which, for clear historical reasons, derives from the doctrine of personality. The question of meaning-fulfilling existential relationships was raised regarding the personality separated from all immediate communal relations. The practical human sciences were developing at the same time that communities were being destroyed; therefore, these practical sciences became the doctrine of personality. A doctrine of community originated only in the forms of a doctrine of the state and a doctrine of the religious community. People forgot that the other communal relationships also represent independent fulfillments of meaning. This situation found formal expression in the classification of the *doctrine of obligation* above the *doctrine of the good*. People spoke of the obligations of individuals within the different relationships of the social life, not seeing that these obligations are rooted in the independent meaning of communal relationships. They forgot that personality becomes personality only by placing itself within meaning-fulfilling existential relationships. The separation and isolation of the personality, followed by the question of its correct conduct toward the community—this is the consequence of an attitude that is opposed to norms. The pure doctrine of personality logically expresses the practical dissolution of real existential relationships.

Ever since romanticism classified the doctrine of the good above the doctrine of obligation, there has been a growing opposition to the one-sided doctrine of personality. It becomes obvious that the separation of the individual from the community is not only nonsensical, but is also an impossible abstraction, and that even the denial of communal relationships is a form of dependence on community. Above all, it became clear that the *life of the community* requires an independent construction of norms that is unrelated to the question of the individual's attitude toward the community. The conflicts between personal claims and professional duty, particularly within representatives of the political and economic communities; insight into the morally destructive effects of certain social situations; the question of the relation between morality and politics—these factors inevitably led to the demand for an independent doctrine of community. The classification of the doctrine of personality within the doctrine of community clearly fulfills this demand.

Of course this raises the danger of a one-sided interpretation. The doctrine of community can be presented so that it neglects personality, just as the doctrine of personality overlooked community. But where this happens, the doctrine of community loses its real meaning, for only communities of personalities can be subject to spiritual norms. The sociological union of subpersonal psychic beings is not community. This justifies the doctrine of personality. There are cases in which the individual personality must free itself from lower existential relationships for the sake of higher ones—not for the sake of permanent isolation, but in order to reestablish adequate communal relationships through a higher communal ethos. The *freedom of the personality* from the community is always freedom for the community. It is freedom from false community for the sake of genuine community.

The object of the doctrine of community is therefore the community of personalities. In every meaning-fulfilling existen-

tial relationship, the doctrine of community treats the nature
of this relationship (the *community*) as well as the bearer of the
relationship (the *personality*). It is never divisible into the doctrine of community and the doctrine of personality as two independent disciplines; every one of its propositions contains
both aspects. It is both possible and necessary to speak of the
development of the personality as such. But as soon as one examines the contents of the personality instead of its form, one is
dealing with the social relationship. Even the human relation
to nature is subject to social formation. When it is not an
aesthetic or metaphysical relation, a relation to nature that excludes the social dimension is a mere existential relation, not a
fulfillment of meaning. Every concept that represents the personality's behavior toward nature according to norms also
describes the socially regulated personality, the personality
that has been freed from its immediate subjection to nature by
means of this normative regulation. Personality is possible only
through community, because freedom is possible only through
community.

If the doctrine of personality is necessarily included in the
doctrine of community, and if the doctrine of personality is
always based on the doctrine of community, either directly or
indirectly, then one cannot forget that the tension between the
elements of meaning is expressed in the *tension between community and personality*. The investigation of this tension is
crucial for the spiritual history of community; the ideal unity
of personality and community is the normative concept that
must be the goal of every doctrine of community. But this
shows that it is impossible to separate the two elements, or to
reduce one of them to the other.

iii. The Doctrine of Community and Ethics

Community is a supported existential relationship. It is supported by ethos, but it itself is not ethos. We have avoided using concepts such as "social ethics" and "individual ethics,"
because we want to avoid the impression that the communal
function is already, in and for itself, ethos. The situation here
is exactly the same as in the theoretical sphere. Science and art

are supported by metaphysics, but they themselves are not metaphysics. It was disastrous for both ethics and the doctrine of community when ethics was reduced to either the doctrine of community or the doctrine of personality. For ethics, the result was the destruction of ethos (i.e., the active direction toward the Unconditioned). Ethics became morality (i.e., the doctrine of dutiful behavior within the domain of the conditioned). Both the Enlightenment's doctrine of obligation and romanticism's doctrine of the good forgot that all social and personal forms can have only symbolic significance for the realization of the Unconditioned. A rational ethics — an *ethics without ethos* — corresponded to rational metaphysics. But just as the metaphysical intention transcends every scientific and aesthetic intention, although it is the basis of both science and art and uses scientific and artistic means of expression, so the ethical intention transcends community and law, although it is the basis of them both and can be realized only in communal and legal forms.

When ethos disappeared from the doctrines of community and personality, both of them lost their import. We have already mentioned the results: community was subjected exclusively to law, and personality was subordinated to the formal demand for universally valid activity. An import-filled doctrine of community is possible only on the ground of an ethic supported by ethos, not on the ground of a *doctrine of morality*. But the relation between ethics and the doctrine of community should not be conceived in such a way that ethics is allowed to heteronomously invade the doctrine of community with special demands. In its form, the doctrine of community is completely autonomous. The import present in the communal forms is determined by the basic ethical attitude, however; this import finds its symbolic expression in the ethical ideal. The treatment of ethics presents the details of this relation. Here it is important to distinguish between ethics and the doctrine of community. Both the dignity of ethics and the independence of the doctrine of community depend on this distinction.

The relationship among ethos, community, and personality can be clarified by reference to the common, though erroneous, classification of duties into *duties toward God, toward one's self, and toward others*. It is irreverent to speak of duties toward God alongside other duties, for there is only one duty, and that is toward God—the duty of direction toward the Unconditioned, the direction that is contained in all activity but that can never be the object of a particular action. Nor are there duties toward one's self alongside duties toward others; there is only the obligatory force of the community of spiritual personalities. Every action must serve this community, whether this action be directed to the actor himself or to others. Duty toward God is the object of ethics; duties toward communities and their representatives are the object of the doctrine of community.

iv. Politics and the Doctrine of the State

The norms of the doctrine of community are realized within social gestalts by means of *politics*. Politics is technological activity within society. It is determined by both the possibilities inherent in the sociological and historical material and the goals derived from the doctrine of community. Politics is a complex science, combining technological and normative elements. Therefore, it has no independent place in the system of the sciences.

Politics can exist in every community, for politics refers not only to the realizations of law, but more generally to the formation of communal law. Thus it also refers to those forms not contained in law. There is politics for both legal and communal forms. Most important, there is politics in the law-bearing community, the state. Politics is instrumental in the formation of the general will; this formation is the presupposition of all realization of law. Because the community with legal power influences the formation of law and thereby the forms of life in all communities, the *politics of the state* is the central political task.

In its principles, the positing of goals in the politics of the state depends on the *normative doctrine of the state* and thus on that aspect of the doctrine of community that is concerned with the state as a community with legal power. The task of the normative doctrine of the state is to show what the law-bearing community should be like. It must show this on the basis of both the philosophy of the state and a description, furnished by the history of spirit, of the different kinds of law-bearing communities. But here the doctrine of the state must bear in mind both the position of the state in the life of the community in general and its central significance for the realization of the idea of community. The tension between the elements of meaning, a tension resulting from the tension between the law-bearing community (the state) and the other immediate forms of community (society), is fundamental for the spiritual history of the state and the doctrine of the norms of the state. The normative concept constructed by the systematics of the state then follows from the ideal balance between the elements in tension.

c. Ethos

Law and community are supported by ethos. The nature of ethics is clarified when ethics is distinguished from the doctrine of community. Ethics is the science of ethos; that is, it is the science of the *active realization of the Unconditioned*. The doctrine of law deals with the meaning-fulfilling existential relationships from the perspective of their rational, thought-determined side, and the doctrine of community deals with them from their irrational, import-determined side; but ethics asks for the import itself, apart from every individual realization.

The place of ethics within the practical sphere corresponds to that of *metaphysics* within the theoretical sphere. Therefore ethics is closely related to metaphysics. Every proposition of a creative metaphysics is an expression of an ethos; every ethos expresses a metaphysics. This is why a metaphysician like Spinoza called his metaphysics *Ethics*.

If ethos is analogous to metaphysics, it is necessarily subject to the same dialectic that every metaphysics is: it attempts to realize the Unconditioned by the formation of existential relationships, that is, by action. Now, an action that is directed toward the Unconditioned can attain its goal only if it itself is unconditioned. But unconditioned action is just as paradoxical as is unconditioned knowledge. Therefore, just as all concepts symbolize the Unconditioned, so all unconditioned existential relationships are symbolic. That which ought to be realized transcends absolutely that which is in fact realized. The con-

cepts by which ethics seeks to grasp the realization of the Unconditioned are to be judged from this standpoint. In their intention, these concepts transcend the proper meaning they
have in the communal sphere. Concepts such as "happiness,"
"wisdom," "the vision of ideas," "humanity," and "the realm of
reason," like corresponding theonomous concepts such as
"blessedness" and "the Kingdom of God," are *transcendent
symbols* for the realization of the Unconditioned. They lose
their meaning as soon as they are drawn down into the sphere
of the conditioned and are made into rational ideas, or
utopias. The practical Unconditioned is misunderstood when it
is made an object of rational action, just as is the theoretical
Unconditioned when it becomes one object of knowledge
alongside other objects. Ethos, like metaphysics, is an attitude.
When it attempts to realize itself, it creates communal and
legal forms that are unconditioned in intention but conditioned in reality. The attempt to realize the unconditioned community as a concrete reality is as contrary to its nature as the
attempt to grasp the Unconditioned as an object of theoretical
knowledge. When this attempt is made, it is justified only insofar as it acknowledges its own symbolism.

The unity of ethics and metaphysics is also apparent in the
fact that both meet in the *metaphysics of history*, where
metaphysics presents the symbols of "is," ethics the symbols of
"ought." The metaphysics of history is therefore that point
within the human sciences in which the theoretical and the
practical are indissolubly connected.

Ethics consists of the philosophy of ethos, the spiritual
history of ethos, and the doctrine of the norms of ethos. The
philosophy of ethos is concerned with the essence and
categories of ethical meaning-giving. The *spiritual history* of
ethos is neither a history of morality nor a history of ethics; it
understands and arranges ethical norms. *Systematic ethics* produces the ethical ideal, which arises from the tention between
the elements of meaning. For this, it uses concepts derived
from the doctrine of community and law. But systematic ethics
does not interfere with either doctrine, any more than
metaphysics interferes with science and art. Because it
recognizes the symbolic nature of these concepts, it cannot
enter the sphere of the conditioned existential relationships

and place its unconditioned norm alongside the conditioned ones. But this is precisely why its attitude can be effective in the comprehension of every individual legal or communal norm. In this sense, but only in this sense, the doctrines of law and community are ethics—in the same sense, therefore, that science and art are metaphysics. This interpretation returns to ethics the dignity it lost when it was combined with the doctrines of law and community. Ethics is concerned with neither the good nor obligation, with neither the personal order nor the legal order. It is not moral philosophy; it is the science of ethos, that is, the science of the realization of the Unconditioned within meaning-fulfilling existential relationships.

C. The Attitude of the Human Sciences

1. The Elements of the Theonomous Human Sciences

a. Theonomy and Autonomy

Theonomy is a turning toward the Unconditioned for the sake of the Unconditioned. The autonomous spiritual attitude is directed toward the conditioned, and toward the Unconditioned only in order to support the conditioned; theonomy employs conditioned forms in order to grasp the Unconditioned in them. *Theonomy and autonomy* are thus not different functions of meaning, but different directions of the same function. They do not stand in a simple opposition, but in a dialectical one: they are based upon the dialectic between the elements of meaning, thought and being. Theonomy is directed toward being as pure import, as the abyss of every thought form. Autonomy is directed toward thought as the bearer of forms and their validity. The tension between the elements of meaning attains its ultimate depth but also its fundamental resolution in the tension between theonomy and autonomy. For it is obvious that all meaning-fulfillment is possible only in the unity of theonomy and autonomy: autonomy alone drives toward empty form without import, and theonomy alone drives toward import without form. But both options are impossible. Living meaning exhibits infinite transitions between the elements of meaning. The ideal meaning of all meaning is the union of both elements.

There are *spiritual situations* in which theonomy predominates and those in which autonomy is dominant. Radical extremes, such as the realization of autonomy in the Greco-Western tradition, are rare. There are usually tensions as the result of the predominance of one of the two elements. The struggle between theonomy and autonomy is the most profound force within the creative spiritual process; this struggle is the dialectical stimulus of history, never allowing history to come to rest. Both theonomy and autonomy are present in every situation, however. Theonomy, the orientation of all forms toward the Unconditioned, can be realized only within forms that are subject to the law of form and thus tend toward autonomy; autonomy cannot be directed to forms without grasping the import they express and thus without the theonomous element.

There is a conflict between the two elements when theonomy sanctifies and preserves forms that contradict the consciousness of validity, and when autonomy rationalizes symbols that express import, either establishing or opposing them. When this happens, theonomy becomes heteronomous: it creates one particular function alongside the others, religion, which, by virtue of its inherent unconditionality, violates and suppresses the others. And autonomy becomes secular: it creates culture, or the totality of meaning-fulfillment outside religion. Thus we have the great, irresolvable *conflicts between religion and culture*, between ecclesiastical and secular metaphysics and ethics, and so on. But this situation is inherently unsound, for it creates two independent functions of meaning from the necessary tension between the elements of meaning, leading to the destruction of both functions. Religion gives unconditioned validity to the unconditioned forms of expression within the autonomous process, and culture rationalizes the symbols of the Unconditioned, depriving them of their meaning and essence.

b. Theonomous Philosophy and History of Spirit

By definition, the *theonomous human sciences* do not constitute a form of science alongside the autonomous ones. When the two coexist, we have an unresolved cultural conflict. This situation confronts both these attitudes with the task of over-coming the conflict: autonomy has the task of absorbing the theonomous element, theonomy, of absorbing the autonomous element. For there is really only one human science, which must proceed from the Unconditioned and from the condition-ed, from import and from form. If our system of the human sciences has acknowledged the duality of spiritual attitudes, this is an expression of the actual situation; but this situation contradicts the ideal state, in which the two are united.

The very *system of the functions of meaning* leads to the conquest of this duality, but at the same time, it shows that the nature of spirit enables this duality to reappear continually. The supporting functions, where the Unconditioned is supposed to be grasped, are impossible within a purely autonomous spiritual situation; the supported functions cannot survive in pure theonomy. Autonomous metaphysics and ethics are ab-sorbed by the supported functions, losing their independent meaning. On the other hand, in pure theonomy the supported functions are heteronomous, losing their validity; indeed, this happens more to the form-determined functions, science and law, than to the import-determined ones, art and community. The point of departure for radical theonomy is therefore always metaphysics and ethics, that of radical autonomy is science and law; art and community are sometimes influenced from the one side, sometimes from the other, and in times of conflict, they often represent the refuge of the spirit.

The task of *theonomous philosophy* is to establish the rela-tion between theonomy and autonomy, both in general and within the individual functions of meaning and their categories. Its highest task is to surrender its own independence and to exhibit its unity with autonomous philosophy. The less it succeeds in this and the more it operates within the cultural conflict, the more it becomes the philosophy of a special fun-ction, the *philosophy of religion*, which stands over against the *philosophy of culture*. The term "philosophy of religion" already symbolizes the state of conflict between autonomy and theonomy. It indicates that religion should be regarded, and defended, as one function alongside the others — an attempt that must necessarily fail, because of its inherent contradictions. This attempt leads to either the rational destruction of religion or the heteronomous destruction of culture. The philosophy of religion has a right to exist only as a theonomous doctrine of the principles of meaning. This doctrine does not stand alongside or above the autonomous doctrine, but together with it conceives the *one* correct doctrine of the principles of meaning. The ultimate goal of the philosophy of religion is self-annihilation in favor of a theonomous philosophy containing autonomy as an equally legitimate element.

The *theonomous history of spirit* understands and arranges the relation between autonomy and theonomy within history. It is not the *spiritual history of religion* — although it is *also* that, just as theonomous philosophy is *also* the philosophy of religion. But it transcends religion, looking for the theonomous import (which cannot be completely lost, even in the most radical autonomy) within every cultural creation. By definition, therefore, the theonomous history of spirit does not stand alongside the *spiritual history of culture*; it is one with this history, and every investigation must overcome this duality of views. This one great path of creative meaning-fulfillment runs through all history, subject to the tension between the elements of meaning. The task of the theonomous history of spirit is, in union with autonomy, to trace this path.

c. Theonomous Systematics (Theology)

One can understand the medieval conception of *theology* as the queen of the sciences only in terms of the relation between the autonomous and theonomous disciplines. Theology was theonomous systematics in general, and until the High Middle Ages it contained autonomy. Only with the decline of the medieval spirit did this unity disintegrate. Theonomy became heteronomous, autonomy became rational. Theology became a particular science alongside of which the autonomous sciences arose. Theology still finds itself in this situation, with all its conflicts. This has become disastrous for it, because it contradicts its nature for theology to be one science alongside the others.

In accordance with the antithesis between rationality and

heteronomy, theology is today treated in either a secular-rational or a religious-heteronomous way. The *secular-rational* conception of theology makes theology the science of religion, of Christianity, of faith, and so on. At the basis of this conception is the truth that God cannot be one object alongside other objects and that there can therefore be no science of God alongside other sciences. The protest against rational metaphysics stands behind liberal theology; but it appears that theology is necessarily abolished in a purely autonomous spiritual attitude. The problem cannot be solved by having some other object replace God as the object of theology, because all other objects are distributed among the autonomous sciences. Theology would have to constitute itself as a practical community of labor concerned with the problems of religion in general and Christianity in particular; but then it would have ceased to exist as an independent discipline.

Opposite this secular-rational conception of theology is the *religious-heteronomous* one, which seeks to retain theology's character as an independent science without having the courage to make it simply the theonomous human science. This heteronomous position attributes absolute significance to the confessional symbols that originated in a theonomous spiritual situation. Then it comes into conflict with the autonomous process in all areas. It retains its metaphysical orientation and understands itself as the science of God, but it forgets that metaphysical symbols depend on the concepts and intuitions produced by the autonomous spiritual process. Thus it makes God one object alongside others; it cannot prevent the transition to rational metaphysics. Whenever it predominates, it divides the consciousness into two kinds of truth, both of which claim to be correct in the same sense. This bifurcation would disrupt the *unity of the spiritual life* if there were no permanent escapes: the sacrifice of autonomy as an act of religious asceticism, in Catholicism; the relegation of religion to the personal, practical sphere, in the Anglo-American world; the concealment of the contradiction by the arts of theological apologetics, in Lutheranism. But because all these means are inadequate, theology always degenerates into a secular-rational science.

Theology is the *theonomous doctrine of the norms of mean-*

ing. This is its only justification. The truth of theology is dependent on the degree to which it abolishes itself as an independent discipline and, in union with autonomous systematics, is the normative human science in general.

d. The Tasks of Theology

Theology is theonomous systematics. This definiton in principle excludes the *empirical view* of religion and of Christianity from theology. Theology is neither the psychology nor the philology nor the history, either of religion in general or of *one* religion. If, following ordinary linguistic practice, one includes these things in theology, then theology is no longer a systematic science; it has become a community of labor. But the criterion for determining the nature of theology cannot be derived from the division of labor within the faculties; it must come from the thing itself. From the perspective of the system of the sciences, theology is theonomous systematics.

Systematic clarification yields a relation of theology to historical investigation, a relation corresponding to that in the area of law. *Historical theology* is not merely philology and history. It goes beyond them in two directions: on the one hand, it is the theonomous history of spirit; on the other hand, it is normative exegesis. As the theonomous history of spirit, it is subject to the demand that it limit itself neither to the history of one particular religion nor to the history of religion in general, but that it take into account the entire process of spiritual realization. As normative exegesis, it is subject to the same problematic as is the doctrine of law. Normative exegesis expresses the creative nature of both effective law and effective theonomous meaning-giving. Theology, too, works on the basis of the concrete spiritual process; it is bound to the classical symbols in which the theonomous conviction has been expressed. So it has a twofold task: to present the original spirit of the religious documents, and to convey this spirit to the contemporary consciousness. As long as the tension between these two is not excessive, the task encounters no basic difficulties. If the tension becomes too great, there are two alternatives. Either the normative document is adjusted to the contemporary consciousness by giving this document a new interpretation—the noble but contradictory attempt of the allegorical method. Or the contemporary consciousness is heteronomously subjected to the legalistically interpreted document of revelation—an attempt that must necessarily fail, leading to a complete alienation from the document. Normative exegesis in the legitimate sense is neither of these two things. The task of normative exegesis disappears and is replaced by a purely historical investigation as soon as the religious consciousness can no longer establish a living relationship with the religious intention of a document of revelation.

Just as law is supported by the state as the law-positing community, so religion is supported by the church as the symbol-creating community. This explains the *concrete, confessional character* of theology. Confessional theology would not be avoided even if all humanity were united in *one* confession; even then, theology would be creative, or confessional. Theology should no more seek a rational religion of reason than the doctrine of law should seek a rational, natural law. Theology can derive its normative system only from the concrete norms of the living confessions. The confessional nature of theology therefore constitutes no objection to its scientific nature. Indeed, this confessional element is essentially bound up with theology as a normative human science. The confessional element is fatal only when it destroys the intention toward the universal, when it misuses the idea of revelation by positing a specific historical symbolism as absolute, thus truncating the creative spiritual process.

In the organization of the theological disciplines, *practical theology* usually stands alongside historical and systematic theology. Our own position yields the following consequences for the systematic classification of practical theology. All the historical material presented by practical theology belongs to the historical sciences, insofar as this material is not used in the history of spirit or normative exegesis. The practical material proper belongs within psychological and sociological technology, where we have already assigned it a special place. But the fundamental theonomous investigations of community, law, art, and science belong to systematic theology. The reason theology no longer pursues these investigations is that theology in general has forfeited its nature as theonomous systematics. The investigation of the objects of theonomous systematics must demonstrate how theology can reestablish itself as theonomous systematics within the individual areas.

2. *The Objects of the Theonomous Human Sciences*

a. Theonomous Metaphysics

The dialectic between theonomy and autonomy is present in all functions of meaning, but it manifests itself differently in the supporting functions than it does in the supported functions. In the supporting functions, the Unconditioned ought to be grasped and realized as the Unconditioned; in the supported functions, the conditioned forms ought to achieve meaning-fulfillment in their independence and validity. This means that a direct *theonomous intention* is possible only in the supporting spheres of meaning and that a direct *autonomous intention* is possible only in the supported spheres. In the first case, autonomy is present only indirectly, in the way symbols are selected; in the second case, theonomy is present only indirectly, in the way forms are grasped. Thus only metaphysics and ethics express theonomy directly.

In the theoretical sphere, the theonomous intention is myth, dogma, or metaphysics, depending on its relation to autonomy. Autonomy is strongest in metaphysics, weakest in pure myth. On the other hand, rational metaphysics attempts to abolish the theonomous intention and to replace it with the autonomous one — an attempt that necessarily fails. Metaphysics is essentially theonomous. Nevertheless, it cannot relinquish the autonomous element. For when autonomous science and art are present in a spiritual situation, metaphysics must derive its symbols from these functions. It cannot remain myth; it must become theonomous metaphysics.

In theology, theonomous metaphysics is usually treated as *dogmatics*. The word "dogmatics" indicates that the symbols it uses are not selected subjectively; they are the decisive expression of a community. Considering what we have said about the concrete, confessional nature of theology, this cannot impair the truth value of dogmatics. That would only be so if the symbols of the community were binding on the dogmatic thinker, in the sense that his goal of knowledge were the correct presentation of dogma, not the truth in dogma. Such a task is both possible and necessary, but its completion belongs to history, not systematics. Theology's presentation of the confessional symbols is usually called "*symbolics*." In its intention, however, normative dogmatics is directed toward the universal. It cooperates in the formation of symbols; it is thus a function of the spiritual life, a function that is necessary as long as there are theonomously fulfilled communities. But dogma can be canonically evaluated in different ways; this can lead to the tragic conflicts we have already mentioned in our discussion of the doctrine of the norms of meaning.

The most immediate form of theonomous metaphysics is the formation of myths; it precedes the emergence of an autonomous science. The most rational form is metaphysics. Metaphysics uses the symbols supplied by an autonomous science. But every *myth*, if it is more than fantasy, contains an autonomous element, a will to grasp the world cognitively; and every *metaphysics*, if it is more than a doctrine of pure form, contains a theonomous element, a mythical will to grasp the Unconditioned. Myth and metaphysics are therefore subject to the same dialectic as are thought and being, form and import. The synthesis of the two is *dogma*, or the attempt to use scientific concepts as theonomous symbols. The danger of this synthesis is that it has neither the immediate freedom of the myth-making consciousness nor the rational infinity of scientific concepts; so it is possible that its rational symbols will be considered unconditional and eternal. When this happens, dogma becomes heteronomous, and the conflict between dogma and metaphysics appears. But when metaphysical symbolism is creatively adapted to both the fundamental theonomous attitude and the autonomous conceptual material, dogmatics accomplishes its synthetic task: it becomes theonomous metaphysics.

b. Science and Art in Theonomy

The supported functions of meaning, science and art, are subject to the direct control of autonomy. All scientific and artistic formations express autonomous meaning-giving. When they do not, they lose their character and become metaphysics. Theonomy can thus be present in science and art only indirectly, in their expression, not in their validity. Theonomy can live in them only insofar as they are supported by metaphysics. The supported nature of science is evident in its methodological attitude; the supported nature of art, in its artistic attitude, or style. There are thus theonomous and autonomous attitudes in both science and art, but in principle there is no *theonomous science and art*, just as there are autonomous and theonomous attitudes in metaphysics, but in principle there is no autonomous metaphysics. The autonomous intention draws metaphysics into the supported sphere of meaning and destroys it; the theonomous intention would transform science and art into metaphysics, thus destroying them.

The implication of this situation for theology is that there can be no *theological aesthetics and doctrine of science*. In relation to the autonomous process of meaning-fulfillment, theology's only task is that of selecting material for theonomous systematics. It can show which attitudes of science and art are theonomous and accordingly, which concepts and intuitions are appropriate for the theonomous formation of symbols. But theology itself cannot create material within the supported spheres. It can influence these spheres only indirectly, by determining the attitude and style of scientific and artistic creations. The clarification of these relationships is exceptionally important, because it prevents the theonomous intention from becoming heteronomous, destroying the supported spheres of meaning. The ideal would be that, on the basis of a theonomous attitude in the supported functions, all scientific and artistic creations could directly serve the formation of symbols of a theonomous metaphysics. But because of the infinity of the rational process, this ideal is unattainable.

If we give the turn toward the Unconditioned within the theoretical sphere the theonomous name "devotion," then the theonomous treatment of science and art yields a theological task whose significance is misunderstood, especially in Protestantism: the *theory of the forms of devotion, or liturgics*. This is a systematic task that is related to the whole of reality, considering it from the point of view of its theonomous power of expression. To consider liturgics as the doctrine of the specifically ecclesiastical forms of devotion is a typically heteronomous constriction; it has ultimately led to assigning liturgics to practical theology. But this classification only applies to its pedagogical, social-technological side, not to its foundation. The theory of the forms of devotion represents the penetration of theonomous metaphysics into the artistic and scientific conception of the world.

c. Theonomous Ethics

From time immemorial, *theological ethics* has existed alongside dogmatics. But though dogmatics had a clear comprehension of its own nature and problems, theological ethics has never been completely clear about itself. The reason for this has been its lack of insight into both the nature of ethos and the relation of ethos to law, community, and personality on the one hand, and to metaphysics on the other. Hence, a substantial part of the ethical material was dealt with in dogmatics, and the remaining part was treated, together with the doctrine of community, in ethics—when the whole of ethics was not included in dogmatics. But as soon as a special theological ethics came into being, the question of its relation to philosophical ethics arose. The common view cannot answer this question. *Theological and philosophical ethics* coexist in an unrelated way. Thus we have a double truth, though it is less sharply perceived than in the theoretical area. This relationship normally expresses a cultural conflict between autonomy and theonomy. This conflict can be overcome only by uniting a theonomous intention with autonomous forms of expression in a theonomous ethos.

Theological ethics is the doctrine of *theonomous ethos*, or the doctrine of piety. The tension between *cultus and ethos* in the doctrine of piety corresponds to the tension between myth and metaphysics. Just as myth precedes the origin of rational science, so the cultus precedes the genesis of a rational consciousness of law and community. And just as metaphysics uses the rational forms of science for theonomous symbolism, so ethos uses the rational forms of law and community for the presentation of piety. But even here the contrast is dialectical. In every cultic act through which the Unconditioned ought to be realized in the existential relationships, there is an element of a rational conception of community and personality; in every ethical act that is still ethos and not the rational formation of being, there is the cultic will to realize the Unconditioned within itself.

The doctrine of *piety* attempts to synthesize cultus and ethos. Like dogma, piety can attempt to heteronomously subjugate law and community. In that case, it leads to the severest conflicts, which destroy community and personality and which express the fact that the synthesis of cultus and ethos is unattainable by theonomous ethos. On the other hand, piety would cease to be ethos if it renounced its theonomous intention and wished merely to create valid forms of community or personality. Piety is meaningfully fulfilled only in the unity of theonomous intention and autonomous form of realization.

d. Community and Law in Theonomy

In the supported functions of community and law, the intention is necessarily autonomous, being directed toward the finite forms of the communal and legal relationships. There is therefore no *theonomous doctrine of community and law*. Theonomous ethos is effective only when it determines the attitude of law and community: just as there is a metaphysical attitude in art and science, so there is an *ethos of law* and an *ethos of community* (including an ethos of personality). Theonomy is present only when law and community are supported by ethos, not in a particular place, but in every form of the supported functions.

This appears to contradict the facts that there are *sacred communities* and *sacred personalities* and that religion creates concrete forms of law, community, and personality. It is important to recognize, however, that these communities and personalities have a paradoxical, figurative meaning, like metaphysical symbols. In intention, sacred communities and personalities are directed toward the Unconditioned, but in reality, they realize conditioned forms. To the extent that religion as a special function forms communities, it creates spiritual communities that are subject to the norms of the spiritual community. The same thing is true of religious personalities who are members of these communities. But there are no unconditioned communities or personalities. The tension between intention and form of expression is present here, exactly as it is in metaphysical symbols, especially in the concept "the Unconditioned." Just as there is no science of the Unconditioned alongside the other sciences, so there is no unconditioned community alongside other communities. When the religious community is interpreted as a special community, it loses its meaning and heteronomously destroys the other communities, instead of giving them theonomous direction. Here, too, the ideal would be that all forms of community and personality would be so filled with theonomous import that they could appear immediately as forms for the realization of a theonomous ethos. But this ideal is utopian.

For the same reason, there can be no theonomous doctrine of law, only a theonomous ethos of law. *Canon law* is either secular law for a spiritual community of great social importance or it symbolizes ideal law in all communities. Canon law, in the sense of a theonomous doctrine of law for a particular religious community, is therefore impossible; attempts to create canon law lead to great conflicts between secular and ecclesiastical legal organization. These conflicts can be resolved only by recognizing the symbolic, paradoxical nature of religious community and by creating, not theonomous law, but an autonomous law supported by theonomous ethos.

For theology, the result of the relation of theonomy to community and law is a *doctrine of the cultic community*, a doctrine corresponding to the doctrine of the forms of devotion. From among the many legal and communal forms, theology selects those that are significant for the presentation of the theonomous ethos. It does not limit itself to specifically ecclesiastical forms, as it did in Protestantism, with the result that it became a part of practical theology and forgot its systematic significance. Instead, theology looks for those forms in all communal and legal relationships that are supported by theonomous ethos; it points to the penetration of theonomous ethos into the entire sphere of meaning-fulfilled existential relationships. But it itself cannot create any forms in these relationships; theonomy can influence law and community only through ethos.

Conclusion

1. Science and Truth

The triad of the sciences of thought, being, and spirit exhausts the system of the sciences. A science outside these three groups is inconceivable, for the nature of knowledge itself is contained in thought, being, and spirit. But to what extent do these three groups, and thus science in general, correspond to the idea of knowledge? That is, what is the *truth character* of science and its parts? This question can be raised only after the entire system is completed. It is the final and most profound question concerning the science of knowledge.

Science is a meaning-fulfilling function. It is a spiritual act, participating in the creative character of everything spiritual. The question concerning the truth of science is therefore answered in principle by means of the notion of creative science. Creation is the realization of the universal within the individual. Science is creative insofar as it realizes the unconditioned form to which it is directed — not as a universal form, but as an individual form. From creativity it follows that science as a whole is convictional, depending on the form of certainty peculiar to the human sciences. But this dependence is different in the different parts of science, and the question concerning the truth of science can be answered only by investigating the influence of the fundamental attitude of the human sciences on the other parts. This procedure also illuminates both the relationships among the three major groups and the living context of the entire system.

The cognitive attitude of the *thought sciences* is intuition, and their degree of certainty is self-evidence. There appears to be no place in these sciences for individual realization and creative conviction. One cannot create such a place even by

217

demonstrating that there have been different forms of logic and mathematics in the history of culture. For example, the appeal to the logic of the primitives, which is completely different from Aristotelian logic, is misleading. Primitive logic is not a scientific comprehension of logical forms; it is the metalogical use of logic in a conception of the world that is naively theonomous and determined by import. Even the propositions of primitive logic contain universal logical axioms. The same thing is true of the mystical, symbolic use of mathematics, a use that can never be completely separated from the foundations of the number system; it is certainly true of the different branches of mathematical science, all of which are mutually related within a context that can be mathematically formulated. The creative element of the thought sciences is not found within its object, but within the *discovery* of its object. From the infinity of formal contexts, those are discovered that conform to the spiritual attitude and spiritual situation. Only those pure forms constituting the meaning-fulfillment in a spiritual situation are newly apprehended in this situation. The other forms are not apprehended. But what is apprehended is universal and self-evident. It cannot lose its self-evidence by any change in the spiritual situation; but it can be seen from another perspective, and it can be absorbed into the formal contexts that have become visible in a new spiritual situation. This is how the thought sciences depend on the attitude of the human sciences. The science of thought is creative in this sense, but only in this sense.

In the *sciences of being*, the empirical sciences, creativity is present in the object itself. This is because of the gestalt knowlege that is fundamental for the empirical sciences. From the abundance of phenomena, gestalt knowledge extracts those elements that constitute the essence of the gestalts. But the essence of gestalts is not only given; it is also a task to be achieved. It cannot just be perceived, it must also be understood. The attitude of gestalt knowledge is perceiving understanding, and its cognitive procedure is re-creative description. But whenever understanding is necessary and re-creation possible, one also finds the attitude and conviction of the human sciences. According to the pragmatic position, knowledge of essence depends on the practical life-contexts within which the knowing and the known gestalts stand. Doubtless these contexts are fundamentally significant for experience; without them, the reality of gestalts would remain hidden. But they have both an inhibiting and a beneficial effect on knowledge. Life-contexts are subjective and accidental; they can create inessential contexts as well as essential ones and can conceal the essence of the gestalt that is to be known. Therefore, the technological-pragmatic relation, for example, from the perspective of which the whole of reality was grasped, produced an enormous amount of empirical material but made a knowledge of essence almost impossible. The creativity of gestalt knowledge is thus not based on the pragmatic, prespiritual, and accidental relation between subject and object, but on the relationship of meaning-fulfillment existing between reality and the spiritual process.

All *meaning-fulfillment* is dependent upon the principles of meaning, the categories. The objects are constituted by the categories, and the methods depend on these objects. The doctrine of the categories is an element of the human sciences. The dependence of gestalt knowledge on creative meaning-giving rests on this doctrine. Gestalt knowledge is creative to the extent that it is determined by the principles of meaning. It is convictional by virtue of the categorial, meaning-giving relation between subject and object. Thus, gestalt knowledge contains two elements: perception (the absorption of phenomena) and understanding (the comprehension of the essences of phenomena). Essence does not consist of the sum of phenomena, though it cannot be apprehended apart from its appearance in the phenomena; essence is apprehended in and through the phenomena by the principles of meaning. There are thus not two acts collaborating in gestalt knowledge; there is only one act, which is both perception and understanding.

Gestalt knowledge is the foundation of all empirical science. Whatever is said about gestalt knowledge is somehow true of all the groups of empirical science, but it is true in these other groups with qualifications and amplifications. In the *physical sphere*, the factor of self-evidence severely restricts conviction. The nearer the physical method stands to mathematics and the less the gestalts are autogenous, the less there is a place for creative realization. There is an analogy between the physical method and mathematics only to the extent that the discovery of methods depends on the attitude of the human sciences. In the physical sphere, therefore, knowledge depends less on categories than on forms. In the *sequence group*, the element of understanding is stronger than that of perception. The understanding of individual spirit-bearing gestalts is not only conditioned by categorial meaning-giving, it also depends on the construction of meaning in general by the history of spirit and systematics. Even the "essence" of the spirit-bearing gestalt is equivocal; in the process of spiritual realization, it continually achieves new fulfillments. This essence is conditioned not only by the empirical situation, but also by the way this situation influences the course of history. A historical gestalt of the past not only *has* an essence, it *fulfills* its essence again and again in the spiritual creations into which it enters and from the perspective of which it is understood anew. Every spiritual creation transforms and fulfills the essence of the entire past. Yet no historical knowledge of essence can abstract from the phenomenal forms of the historical gestalt. Here, too, the knowledge of essence is indissolubly connected with empirical perception. The type of certainty in all empirical cognition is an inseparable unity of probability and conviction.

The creative nature of science is purely expressed in the *human sciences*. Here creativity is not limited by the rational-deductive or the rational-empirical element. The human sciences are productive; they help create the object they know. They participate in the creative spiritual process, and therefore their only limit is the spiritual process itself. The dependence of the human sciences on the spiritual process (including the scientific process) is manifest in the two elements of systematics, the doctrine of the principles of meaning and the doctrine of the material of meaning. In the former, the meaning-giving principles are separated from a creative reality of meaning already present; in the latter, the norms of meaning that have shaped consciousness are elevated to the level of scientific awareness. Thus, both are dependent on the reality of meaning and on a meaning-fulfilled consciousness. But this dependence is not expressed rationally. It influences cognition in the human sciences in an *immediate* way. Every philosophical or humanistic system is convictional. Unlike the theories in the sciences of thought and being, these systems cannot be altered by rational or empirical insights.

The creative nature of science within the various groups is apparent in the *kind of criticism* employed in each group. In the thought sciences, a twofold criticism is possible: one from the perspective of logical consistency, the other from that of the goal that is posited for the discovery of form. There are errors in deduction, and there are unattained goals.

Both forms of criticism are present in the empirical sciences, to the extent that the thought sciences are operative in them. But because of the empirical material, there is also another form of criticism. Every demonstration of an error and every new discovery are of fundamental importance for the knowledge of being. But there is no rational or empirical criticism of the knowledge of essence in itself. For a judgment about what is essential is based upon understanding and is thus convictional.

Formal, logical criticism is still present in the human sciences. This criticism is the most universal criterion of the sciences generally, though it is completely devoid of content. Whatever is logically inconsistent cannot claim to be valid. No other rational or empirical criticism is applicable to the creative system. The actual criticism of the system proceeds from the spiritual process itself, which either absorbs the system or disregards it as insignificant. Whether positive or negative, this criticism is more penetrating than rational criticism. But it itself is not rational. Indeed, it can be expressed in other systems; one can demonstrate the limits of a conviction. But this demonstration is not rational; it is based on another kind of productive comprehension of essence. When it is not formal, criticism in the human sciences is itself creation.

The conception of science as a creative activity solves the problem of *value-free science*. The dispute over the antithesis between valuing and value-free science has not been resolved, because it has not posed the problem separately for the three groups of science. No one saw that the solution to this problem must be different in each area. The creativity characteristic of the human sciences is present in the empirical sciencs only in their comprehension of essence and in the thought sciences only in their discovery of form. Any answer to the question of value-free science that disregards these differences is necessarily misleading.

The unity of science is not destroyed by the recognition that there are differences among the truth characters of the kinds of science. Both the whole of science and every spiritual function are conditioned by the direction toward the universal, on the one hand, and by the creative substance in which the universal is individually realized, on the other. But the direction toward the universal varies according to the differences among the objects of knowledge. In the thought sciences, the conditioned forms are the objects of knowledge; in the human sciences, the principles and norms of meaning-fulfillment are the objects of knowledge. Thus existents as well as valid forms and norms are objects of scientific meaning-fulfillment. This is the paradox of the sciences of thought and spirit. These sciences make that an object through which the object is constituted as meaning-fulfilled. This explains why creativity is not characteristic of knowledge in the thought sciences and why it is the dominant feature of knowledge in the human sciences. But both the thought and the human sciences are elements of the *one* science, which manifests its dual nature perfectly in the empirical sciences.

The dispute over value-free science was necessarily fruitless for another reason. Because of its subjective connotation, the concept "value" itself prevented the antithesis between valuing and value-free science from being clarified and resolved. *Freedom from values* appears to be completely justified as the ideal of every science if valuation is combined with subjective prejudice. If valuation is a source of error, the truth character of a science depends on whether one can eliminate this source of error. But the theory of the creative character of truth overcomes the subjective conception of value. Valuation should mean that all meaning-fulfillment is creative and that science, as meaning-fulfillment, must therefore have a creative form.

This conception contains the *dynamic notion of truth*. Truth is neither the picture of a static reality, as realism thinks, nor the affirmation of established, ideal norms, as idealism supposes. Both of these conceptions of truth necessarily lead to skepticism and relativism. Relativism is the consequence of a notion of truth that makes knowledge passive in relation to the object of knowledge. If the object is absolute, as in both idealism and realism, knowledge is relative, unless there is an intuitive relation between subject and object, an intuitive relation that creates immediate self-evidence. But this is so only in the sciences of thought. The theory of creative meaning-fulfillment is not relativism; it is the conquest of relativism. Truth is the living process of individual, creative meaning-fulfillment, the process in which the object achieves fulfillment in the spiritual act. Thus the theory of the creative nature of science finds its ultimate, decisive expression in the dynamic notion of truth.

2. Science and Life

If science is an act of creative meaning-fulfillment, then it is an act of life itself. The problem of *science and life* is solved, in principle, with the solution of the problem of value-free science.

For rationalism, knowledge is the abolition of the relation between life and things and the establishment of a relation of pure form. It rejects the question of the meaning of knowledge for life; the meaning of knowledge is knowledge itself. Science is purely self-sufficient; it requires no justification by life. It appears to be defiled if it is given a purpose for life. This position recognizes that every meaning-fulfillment is directed toward the unconditioned form and is therefore unconditionally subject to the demand of validity. A science in which cognitions are influenced by heterogenous influences from existential relationships instead of by norms is actually impure and intractably opposed to the unconditioned form. But the truth character of science does not preclude its meaning for life. Rationalism is justified only in its opposition to a heteronomous turning of life against thought, not in its answer to the question of the meaning of science in general for life.

Whenever the rational attitude toward science is assumed, we have the principle of the *contradiction between life and knowledge*, the pessimism concerning science that today leads to a growing alienation from science. But this principle wrongly presupposes that the rational attitude is necessary. If this were so, then the antithesis between science and life would in fact be insurmountable. The pure rational form would desiccate life, leaving empty universal forms, on the one hand, and lifeless material for knowledge, on the other. Whatever would be known would be destroyed in the very knowing.

The *pragmatic conception* of science depends on the rational cognitive attitude, even though this conception opposes the notion of truth possessed by such an attitude. The pragmatic view wants to dissolve knowledge into life; it is unaware of the contradiction between life and thought, for it denies the one side. It claims that all formulations of concepts depend on prespiritual existential relationships. This view does not see that it thereby invalidates itself, transforming itself into a pragmatic assertion, that is, an assertion with no apparent claim to objective validity. Pragmatism is the attempt to avoid the consequences of the rational cognitive attitude without abandoning the presupposition of this attitude. Basically, it proclaims the rational notion of truth with irrational symptoms. It therefore does not provide an escape from the conflict between life and knowledge. The pragmatic notion of truth and the pragmatic attitude toward things obviously deprive these things of life and import. Pragmatism destroys things in order to use them.

The doctrine of the *creative nature of science* avoids the consequences that are adverse to life. For this doctrine, things retain their distinctive life, their import. But the essence of reality is fulfilled in the creative act. Knowledge is a form of meaning-fulfillment toward which everything real is inclined. Knowledge is a co-creation and a continuation of creative positing. It does not violate things; it brings them to fulfillment. But it can be meaning-fulfillment only because it both recognizes the distinctive forms of things and subjects itself to the unconditioned form.

The meaning of knowledge for life in relation to things is manifest in the meaning-fulfilling nature of knowledge; its meaning for life from the perspective of the subject is manifest in its *metalogical form*, that is, in the living reciprocal influence between knowledge and all spiritual functions. All the functions of meaning are present in the metalogical comprehension of being, and the whole spiritual attitude influences the giving of logical form. In metalogic, it is obvious that every act of knowledge is also an act of life, that the cognitive act expresses a relationship between reality and the spirit-bearing gestalt as a whole. This is not to say that the other functions of meaning would be present, as functions, in metalogical knowledge. The only thing present is the comprehension of being, the basic creative attitude that is expressed equally in all the spiritual functions. Metalogic does not violate the autonomy of scientific form, yet metalogical knowledge is an act that creates a living relation to reality.

The same thing is true of the influence of knowledge on the other spiritual functions. The *productive character* of the human sciences reveals a living mutual relationship between the spiritual process and the knowledge of spirit. Knowledge does not provide laws that the other functions of meaning must rationally obey; nor do the functions of meaning have an objective reality apart from knowledge, so that knowledge must be directed to them in an empirical way. Rather, the cognitions of the human sciences are productive only when they are created from the living, autonomous process of the other spiritual functions. But when they are productive, they decisively influence the further formation of the spiritual process — not by means of a heteronomous violation of the autonomous forms of the other functions of meaning from the perspective of scientific form, but by means of the living reciprocal influence between immediate creativity and the creative development of consciousness.

The most profound expression for the meaning of science for life is the fact that science *is supported* by metaphysics. This places science in a direct relation to the unconditioned meaning that gives every individual meaning both significance and reality. In providing the elements of meaning, metaphysics gives science the living, tension-rich principle by which science can grasp the import of reality. Metaphysics extends into all areas of being, indicating in them the unconditioned import expressed in every individual import. Thus the relationship of science to its objects is theonomous. The Eros toward the Unconditioned constitutes the depth of every cognitive relation to reality. All science can serve the Unconditioned; it can be the will to apprehend the Unconditioned in every individual.

Therefore, science does not become metaphysics; it must refuse to be the *doctrine of world views* and to confer the appearance of scientific validity upon metaphysical symbols. Science cannot satisfy the demand to be the scientific doctrine of world views, the demand often placed on it by the contemporary philosophy of life. Whenever it attempts to do so, it reduces the Unconditioned to the level of the conditioned, losing its peculiar dignity, its character of validity. To be sure, it provides concepts to metaphysics; but within metaphysics, these concepts are not scientific, they are symbolic. At the present time, nothing is more dangerous for the reunion of science and life than the attempt to make science the doctrine of world views. Metaphysics influences science only by its entire attitude, and science influences metaphysics only by providing symbols. That is the true, theonomous meaning of science for life, uniting theonomy with autonomy.

In the systematics of the human sciences, we have proceeded from the theoretical and practical acts of the spirit-bearing gestalt. As long as the intention is directed toward the particular forms of both functions, the distinction between these two is fundamental. But as soon as spirit is directed toward the living import of them both, the distinction loses its importance. We have already shown this in the relationship between metaphysics and ethics. These two disciplines are united in the total attitude of the consciousness that is directed toward the Unconditioned. This is also true, in a derivative way, for the relationship between *science and community*. Characteristically, Greek philosophy found concepts expressing the unity of cognitive attitude and community just when it was directed most strongly to the comprehension of import. Plato uses the concept "Eros," from the communal sphere, to characterize the most profound essence of the will to knowledge; in the religious philosophy of Hellenism, the term "gnosis," which is derived from the sphere of knowledge, designates the inner essential union with the object, whether it be a conditioned object or the Unconditioned itself. The meaning of knowledge for life has achieved its highest expression in the union of knowledge and love.

Love is not the annihilation, but the affirmation of the individual form, of the other. Essential love is inherent in justice.

In the sphere of knowledge, justice is both recognition of the individual forms of objects and obedience to the unconditioned form upon which every act of knowledge is based. The meaning of science for life may be proclaimed, without endangering the seriousness and veracity of science, only when the theonomous attitude, the union of knowledge and love, is supported by obedience to the autonomous forms. Like every meaning-fulfilling act, science attains truth only in the perfect *union of theonomy and autonomy.*

Notes

Translator's Introduction

1. From "Connoisseur of Chaos," in *The Collected Poems of Wallace Stevens*, p. 215© 1977 by Alfred A. Knopf. Reprinted with permission.
2. The exceptions are, of course, significant. One thinks of James Luther Adams, *Paul Tillich's Philosophy of Culture, Science, and Religion* (New York: Harper & Row, 1965); Robert P. Scharlemann, *Reflection and Doubt in the Thought of Paul Tillich* (New Haven: Yale University Press, 1969); Clark A. Kucheman, "Religion, Culture, and Religious Socialism," *Journal of Religion* 52 (1972): 268-86; and Victor Nuovo, "On Revising Tillich: An Essay on the Principles of Theology," in John J. Carey, ed., *Kairos and Logos: Studies in the Roots and Implications of Tillich's Theology* (Cambridge, Mass.: The North American Paul Tillich Society, 1978), pp. 45-73. Presumably volume 2 of Wilhelm and Marion Pauck's *Paul Tillich: His Life and Thought* will deal with the work, too.
3. See my article "From System to Systematics: The Origin of Tillich's Theology," in Carey, *Kairos and Logos*, pp. 121-34.

General Foundation

1. See the "Foundation" of the human sciences, pp. 137-58.
2. For a discussion of the method of the following deduction, see section 4 of this chapter, pp. 39-41.

Part 1: The Sciences of Thought (The Ideal Sciences)

1. The concept *"group,"* in the thought sciences, would correspond to the concept "species" found in the sciences of being.
2. Here we cannot discuss the innumerable problems involved with these methodological concepts. We consider methods and objects only insofar as is necessary for the foundation and elucidation of the general structure of the entire system.

Part 2: The Sciences of Being (The Empirical Sciences)

1. It may not be feasible to restore the original meaning of the term "sequence" [*Folge*], which was divested of its temporal meaning and came to be used only for the logical relation of inference [*Folge*]. But no other word expresses the historical context with the same precision with which "law" expresses the physical context and "gestalt" expresses the organic one. Like the thing itself, the linguistic expression for it must be wrested from the dominance of rational, causal thought.

2. We refer to the theory of the categories only insofar as it is significant for the structure of the system.

3. A metalogical theory of the categories is urgently needed. It would be the decisive blow against neo-Kantian logism, whose theory of the categories is completely dominated by mathematical physics. It would also overcome the noncategorial thought of phenomenology.

4. I am indebted to numerous discussions with Dr. Alexander Rüstow in Berlin for my use of the word "gestalt" to refer to a system of self-contained causality. Dr. Rüstow himself used the word in more of a phenomenological sense.

Part 3: The Sciences of Spirit (The Normative Sciences)

1. *On a misuse of the concept "the human sciences"*: In the present situation, the position that the human sciences are creative must be defended against a view that claims for itself the name "the human sciences," or "the sciences of spirit," but understands by this name something entirely different from what we understand by it. This position is that of *anthroposophy*. When it speaks of spirit, it refers to a reality that is a higher form of a psychological gestalt, a reality that is empirically established according to special methods and with special instruments. Now, there is no compelling reason to deny the existence of such a reality and of the instruments of perception appropriate to it. It is even conceivable that in the evolution of organic gestalts, such instruments could become a common possession. But we must emphatically deny that the cognition of this reality deserves the name "the human science." There would have to be an epistemology and a doctrine of categories, an aesthetics and a metaphysics, appropriate to these objects; there would also have to be a social, an ethical, and a religious approach to them. That is, after these objects were empirically known, they would have to be classified by the human sciences according to their meaning and validity. Only this classification would be the human science, the science of spirit. The anthroposophical position misunderstands the nature of spirit as completely as does the heterogenous use of the method of physical law it opposes.

2. Our demand for a metalogical method is not an abstract demand. This nomenclature (which can perhaps be challenged) refers to a method I used, long before I gave it that name and long before I was conscious of its

methodological significance, in my lectures on the philosophy of religion, the history of spirit, and social philosophy. I used the term in my essays on social philosophy entitled *Mass and Spirit*, and in the two essays "On the Idea of a Theology of Culture" and "The Conquest of the Concept of Religion in the Philosophy of Religion." Troeltsch uses the concept "metalogical" in his *Historicism and its Problems* for what I call the "goal of knowledge." The relation between the doctrine of the goals of knowledge and the metalogical doctrine of categories is obvious. Yet is seems to me that the concept "metalogical" should be used, not for a subdivision of the doctrine of methods, but for the type of method itself.

    3. *Translator's note*. Here Tillich inexplicably omits a section on the spiritual history of science. This omission appeared in the first publication of the book.

Index